RENEWALS: 691-4574
DATE DUE

APPLIED CHEMICAL AND ISOTOPIC GROUNDWATER HYDROLOGY

APPLIED CHEMICAL AND ISOTOPIC GROUNDWATER HYDROLOGY

Emanuel Mazor

HALSTED PRESS a division of
John Wiley & Sons
New York   Toronto

Published in the USA, Canada and
Latin America by Halsted Press, a Division
of John Wiley & Sons, Inc., New York

First Published 1991

Copyright © E. Mazor 1991

All rights reserved. No part of this publication may be
reproduced, stored in a retrieval system or transmitted in
any form or by any means, without written permission from the
publisher.

Library of Congress Cataloging-in-Publication Data

Mazor, 'Imanuél.
  Hydrochemical fieldwork/Emanuel Mazor.
    p. cm.
  Includes bibliographical references.
  ISBN 0-470-21652-2
  1. Water chemistry--Field work. I. Title.
GB855.M29 1990
551.48--dc20   89-48327
                                      CIP

Printed in Great Britain

Library
University of Texas
at San Antonio

CONTENTS

Preface ix

1 INTRODUCTION 1

    1.1 Water composition 1
    1.2 Information encoded into water during the hydrological cycle 2
    1.3 Questions asked 4
    1.4 Conceptual models 6
    1.5 Economic aspects of field hydrochemistry 6

2 BASIC HYDROLOGICAL CONCEPTS 9

    2.1 The aerated intake zone 9
    2.2 The saturated zone and the water table 11
    2.3 Aquifers 12
    2.4 Aquicludes 13
    2.5 Phreatic, or free surface, aquifers 14
    2.6 Confined aquifers and artesian flow 15
    2.7 Karstic systems 16
    2.8 Velocity of groundwater flow in the saturated zone 18
    2.9 Modes of groundwater flow in the saturated zone 19
    2.10 Evapotranspiration 20
    2.11 Recharge 21
    2.12 Discharge 22

3 GEOLOGICAL DATA 23

    3.1 Lithology and its bearing on water composition 23
    3.2 Properties of geological materials and their bearing on recharge and groundwater storage 25
    3.3 Layering and its effect on groundwater flow 27
    3.4 Folded structures and their bearing on flow direction and confinement 27
    3.5 Faults controlling groundwater flow 29
    3.6 Intrusive bodies influencing groundwater flow 30

4	**PHYSICAL PARAMETERS**	32
	4.1 Water table measurements	32
	4.2 Interpretation of water table data	34
	4.3 Gradient and flow direction	39
	4.4 The need for complementary data to check deduced gradients and flow directions	40
	4.5 Velocities and pumping tests	42
	4.6 Chemical and physical measurements during pumping tests	42
	4.7 Temperature measurements	45
	4.8 Tracing groundwater by temperature: a few case studies of the Mohawk River	48
	4.9 Cold and hot groundwater systems	50
	4.10 Discharge measurements and their interpretative value	53
5	**ELEMENTS, ISOTOPES, IONS, UNITS, ERRORS**	57
	5.1 Elements	57
	5.2 Isotopes	57
	5.3 Atomic weight	59
	5.4 Ions and valencies	59
	5.5 Ionic compounds	60
	5.6 Concentration units	61
	5.7 Reproducibility, accuracy, resolution and limit of detection	63
	5.8 Errors and significant figures	64
	5.9 Checking the laboratory	65
	5.10 Evaluation of data quality by data processing techniques	66
	5.11 Putting life into a dry table	66
	5.12 Evaluation of calculated reproducibilities and reaction errors	70
6	**CHEMICAL PARAMETERS — DATA PROCESSING**	72
	6.1 Data tables	72
	6.2 Fingerprint diagrams	74
	6.3 Composition diagrams	79
	6.4 Major patterns seen in composition diagrams	80
	6.5 Establishing hydraulic interconnections	82
	6.6 Mixing patterns	85
	6.7 End member properties and mixing percentages	89
	6.8 Water–rock interactions and types of rocks passed	92
	6.9 Water composition	93
	6.10 Compositional time variations	95
	6.11 Some case studies	96
7	**PLANNING HYDROCHEMICAL STUDIES**	102
	7.1 Representative samples	102
	7.2 Data collection during drilling	104
	7.3 Depth profiles	105
	7.4 Data collection during pumping tests	106

Contents   vii

	7.5	Importance of historical data	107
	7.6	Repeated observations or time-data series.	108
	7.7	Search for meaningful parameters	109
	7.8	Sampling for contour maps	110
	7.9	Sampling along transects	110
	7.10	Reconnaissance studies	111
	7.11	Detailed studies	111
	7.12	Summary: a planning list	111
8	CHEMICAL PARAMETERS – FIELD WORK		113
	8.1	Field measurements	113
	8.2	Smell and taste	114
	8.3	Temperature	115
	8.4	Electrical conductance	115
	8.5	pH	116
	8.6	Dissolved oxygen	117
	8.7	Alkalinity	118
	8.8	Sampling for dissolved ion analyses	119
	8.9	Sampling for isotopic measurements	119
	8.10	Preservation of samples	120
	8.11	Efflorescences	120
	8.12	Equipment list for field work	121
9	STABLE HYDROGEN AND OXYGEN ISOTOPES		122
	9.1	Isotopic composition of water molecules	122
	9.2	Units of isotopic composition of water	122
	9.3	Isotopic fractionation during evaporation and some hydrological applications	123
	9.4	The meteoric isotope line	125
	9.5	Temperature effect	129
	9.6	Amount effect	130
	9.7	Continental effect	132
	9.8	Altitude effect	133
	9.9	Tracing groundwater with deuterium and oxygen-18: local studies	138
	9.10	Tracing groundwater with deuterium and oxygen-18: a regional study	142
	9.11	The need for multisampling	145
10	TRITIUM		147
	10.1	The radioactive heavy hydrogen isotope	147
	10.2	Natural tritium production	147
	10.3	Manmade tritium inputs	148
	10.4	Tritium as a short-term age indicator	151
	10.5	Tritium as a tracer of recharge and piston flow: observations in wells	153
	10.6	The special role of tritium in tracing intermixing of old and recent waters	159

	10.7 Tritium, dissolved ions and stable isotopes as tracers for rapid discharge along fractures: the Mont Blanc tunnel case study	160
11	RADIOCARBON AND CARBON-13	164
	11.1 The isotopes of carbon	164
	11.2 Natural ^{14}C production	164
	11.3 Manmade carbon-14 dilution and addition	165
	11.4 Carbon-14 in groundwater: an introduction to dating of groundwater	166
	11.5 Lowering of ^{14}C content by reactions with rocks	169
	11.6 Carbon-13 abundances, and their relevance to ^{14}C dating of groundwater	171
	11.7 Application of $\delta^{13}C$ to correct observed ^{14}C values for changes caused by interactions with carbonate rocks	173
	11.8 Direction of down-gradient flow and groundwater age, studied by ^{14}C: case studies	174
	11.9 Flow discontinuities between adjacent phreatic and confined aquifers, indicated by ^{14}C and other parameters	184
	11.10 Mixing of groundwaters, revealed by joint interpretation of tritium and ^{14}C data	193
	11.11 Piston flow versus karstic flow, revealed by ^{14}C data	197
12	NOBLE GASES	198
	12.1 Rare, inert, or noble?	198
	12.2 Atmospheric noble gases	199
	12.3 Groundwater as a closed system for atmospheric noble gases	200
	12.4 Studies on retention of atmospheric noble gases in groundwater systems: cold springs	201
	12.5 Further checks on atmospheric noble gas retention: warm springs	205
	12.6 Computation of paleotemperatures	208
	12.7 Calculation of depth of circulation and location of recharge zone	211
	12.8 Identifying karstic recharge	213
	12.9 Radiogenic He and its use as an age indicator	214
	12.10 Sample collection for noble gas measurements, and contact with relevant laboratories	216
13	MONITORING OF CONTAMINANTS	217
	13.1 Scope of the problem	217
	13.2 Detection and monitoring of pollutants: some basic rules	218
	13.3 Groundwater pollution case studies	220
	13.4 Nuclear waste disposal	253
	13.5 Summary of case studies	254
14	HYDROCHEMISTS' REPORTS	255
	14.1 Why reports?	255
	14.2 Types of report	255
	14.3 Internal structure of reports	256
	References	264
	Index	271

PREFACE

Classical hydrology deals with aquifer properties and water levels, and from these parameters the flow of groundwater is computed (related textbooks: Freeze and Cherry (1979), Todd (1980), de Marsily (1986)). In the present book emphasis is placed on water properties – physical, chemical and isotopic, and their variations with time. These parameters are applied to understand water systems.

The book is written with the assumption that most field hydrochemists bring their samples to specialized laboratories for analyses and therefore the laboratory work is not dealt with here.

Field hydrochemists plan hydrochemical studies, conduct field measurements and observations, collect samples for laboratory analyses, process the data, formulate conceptual and mathematical models, and make recommendations. This book deals with these topics, emphasizes the processing of data and discusses a large number of published case studies.

A special word of thanks for encouragement and help in the early stages of writing of this book is due to Professor S.N. Davies, Department of Hydrology and Water Resources, University of Arizona, Tucson. Comments and suggestions by Dr M. De Freitas, Engineering Geology Division, Imperial College, London, are warmly acknowledged.

1 INTRODUCTION

1.1 Water composition

Water molecules are built of two elements – hydrogen, H, and oxygen, O. The general formula is H_2O. Different varieties of atoms with different masses, called isotopes, are present in water: light hydrogen, ^1H (most common); heavy hydrogen called deuterium, ^2H or D (rare); light oxygen, ^{16}O (common); ^{17}O (very rare), and heavy oxygen, ^{18}O (rare). Thus, several types of water molecules occur, the most important being:

$H_2^{16}O$ – light, most common water molecule;
$HD^{16}O$ and $H_2^{18}O$ – heavy, rare water molecules.

The major terrestrial water reservoirs, the oceans, are well mixed and of rather uniform isotopic composition. Upon evaporation an isotopic separation occurs, because the light water molecules are more readily evaporated. Thus, water in clouds is isotopically light compared to ocean water. Upon condensation, heavier water molecules condense more readily, causing a 'reversed' fractionation. The degree of isotopic fractionation depends on the ambient temperature and other factors which are discussed in sections 9.5 to 9.8.

Water contains dissolved salts, dissociated into cations (positively charged ions) and anions (negatively charged ions). The most common dissolved ions are sodium (Na^+), calcium (Ca^{2+}), magnesium (Mg^{2+}) and potassium (K^+). Most common anions are chloride (Cl^-), bicarbonate (HCO_3^-) and sulphate (SO_4^{2-}), discussed in sections 5.1 and 5.5. The composition of water, i.e. the concentration of the different ions, varies over a wide range of values.

Water also contains dissolved gases. A major source is air, contributing nitrogen (N_2), oxygen (O_2) and noble gases – helium (He), neon (Ne), argon (Ar), krypton (Kr) and xenon (Xe). Other gases, of biogenic origin, are added to the water in the ground, e.g. carbon dioxide (CO_2), nitric oxide (NO_2), methane (CH_4) and hydrogen sulphide (H_2S). Biogenic processes may consume the dissolved oxygen.

Cosmic rays interact with the upper atmosphere and produce two radioactive isotopes: ^{3}H, better known as tritium (T), the heaviest hydrogen isotope, and carbon-14 (^{14}C), the heaviest carbon isotope. The rate of radioactive decay is expressed in terms of half-life, i.e. the time required for atoms in a given reservoir to decay to half their initial number. The half-life of tritium is 12.3 years and that of carbon-14 is 5730 years. Both tritium and carbon-14 are incorporated into groundwater. They decay in the saturated zone (section 2.2), providing two semi-quantitative dating tools – one for short periods and one for long periods. Manmade tritium and carbon-14, introduced by nuclear bomb tests, complicate the picture, but provide additional information (sections 10.4, 11.4).

Rocks contain uranium and thorium in small concentrations, and their radioactive decay results in the production of radiogenic helium (^{4}He). The He reaches groundwater and is dissolved and stored. With time, radiogenic helium is accumulated in groundwater, providing an independent semi-quantitative dating method (section 12.9), useful for ages beyond the carbon-14 method.

1.2 Information encoded into water during the hydrological cycle

The hydrological cycle is well known in its general outlines: ocean water is evaporated, forming clouds that are blown into the continent, where they gradually condense and rain out. On hitting the ground, part of the rain is turned into runoff, flowing back into the ocean, part is returned into the atmosphere by evapotranspiration, and the rest infiltrates and joins groundwater reservoirs and ultimately returns to the ocean as well. Hydrochemical studies reveal an ever-growing amount of information that is encoded into water during this cycle. It is up to the hydrochemist to decipher this information, and to translate into terms usable by water management personnel (Fig. 1.1.).

Rain-producing air masses are formed over different regions of the oceans, where different ambient temperatures prevail, resulting in different degrees of isotopic fractionation during evaporation and cloud formation. This explains observed variations in the isotopic composition of rains reaching a region from different directions or at different seasons. In terms of the present chapter, we can say that water is encoded with isotopic information even in the first stages of the water cycle.

Wind carries droplets of sea water, which dry and form salt grains. These windborne salts, or sea spray, reach the clouds and are carried along with them.

Rain water dissolves atmospheric N_2 and O_2 and the rare gases, as well as CO_2. In addition, radioactive tritium and carbon-14 are incorporated.

Manmade pollutants reach rain water through the air. The best known is acid rain, which contains sulphur compounds. A large variety of other pollutants of urban life and industry is lifted into the atmosphere and then washed down by rain. As clouds move inland, or rise up mountains, their

Introduction

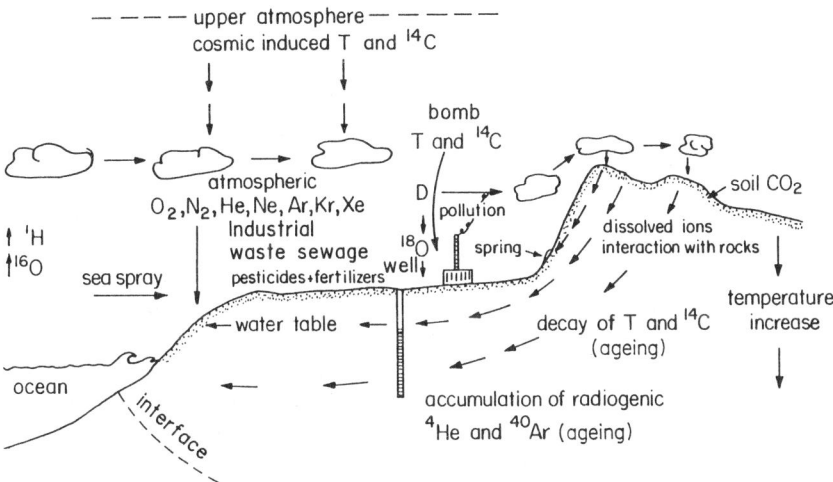

Fig. 1.1 Information coded into water along its surface and underground path: right at the beginning, water in clouds is tagged by an enrichment of light hydrogen (^{1}H) and oxygen (^{16}O) isotopes, separated during evaporation. Cloud water equilibrates with the atmosphere, dissolving radioactive tritium (T) and radiocarbon (^{14}C), produced by nature and introduced by nuclear bomb tests. Also, atmospheric gases are dissolved, the most important being oxygen (O_2), nitrogen (N_2) and the noble gases (He, Ne, Ar, Kr, and Xe). Gas dissolution is dependent on temperature and altitude (pressure). Urban and industrial pollutants reach clouds as well, besides natural sea spray. As clouds produce rain, heavy hydrogen (D) and heavy oxygen (^{18}O) isotopes are preferentially enriched, leaving depleted water in the cloud. Thus, inland and in mountainous regions rains of different isotopic compositions are formed. On hitting the ground, rain water dissolves accumulated dust particles, pesticides and fertilizers. These are carried to rivers or introduced into the ground with infiltrating water. Sewage and industrial wastes are introduced into the ground as well, mixing with groundwater. Underground, water is disconnected from new supplies of tritium and ^{14}C, and the original amounts decay with time. On the other hand, radiogenic helium (^{4}He) and argon (^{40}Ar), produced in rocks, enter the water and accumulate with time. As water enters the soil zone it becomes enriched in soil CO_2 produced by biogenic activity, turning the water into an acid that interacts with rocks, introducing dissolved ions into the water. The deeper water circulates the warmer it gets, due to the existing geothermal gradient.

water condenses and gradually rains out. Isotopic fractionation increases the deuterium and oxygen-18 content of the condensed rain, depleting the concentration of these isotopes in the vapour remaining in the cloud. As the cloud moves on into the continent or up a mountain, the rain produced becomes progressively depleted in the heavy isotopes. Observations of this type have been developed into a source of information on distance of recharge from the coast and recharge altitude, parameters that help to identify recharge areas (sections 9.7 and 9.8).

Upon contact with the soil, rain water dissolves accumulated sea spray and dust, as well as fertilizers and pesticides. Fluid wastes are locally added, providing further tagging of water.

Infiltrating water, passing the soil zone, becomes loaded with CO_2 formed in biogenic processes. The water turns into a weak acid that dissolves soil components and rocks. The nature of these dissolution processes varies with soil and rock types, climate, and drainage conditions. Dissolution ceases when ionic saturation is reached, but exchange reactions may continue.

Daily temperature fluctuations are averaged out in most climates at a depth of 40 cm below the surface, and seasonal temperature variations are averaged out at a depth of 10–15m. At this depth the pevailing temperature is commonly close to the average annual value. Temperature of shallow groundwater equalling that of the rainy season indicates rapid recharge, through conduits. In contrast, if the temperature of shallow groundwater equals the average annual temperature this indicates retardation in the soil and aerated zone. Along its path underground water is heated by the local heat gradient, reflecting depth of circulation (sections 4.7–4.9).

When water is stored underground, the radioactive isotopes tritium and carbon-14 decay and the concentrations left indicate the age of the water. At the same time, radiogenic He accumulates, providing another age indicator. A summary of the information encoded into water during its cylce is given in Fig. 1.1.

1.3 Questions asked

Many stages of the water cycle are described by specific information implanted into surface water and groundwater. Yet field hydrochemists have limited access to the water, being able to measure and sample it only at single points – wells and springs. Their task is to reconstruct the complete water history. A list of pertinent topics is given below.

Water quality. Because groundwaters differ from one place to another in the concentration of their dissolved salts and gases, it is obvious that they also differ in quality. Water of high quality (low salt content, good taste) should be saved for drinking, irrigation, and a number of specific industries. Poorer waters (more salts, poorer taste) may be used, in order of quality, for domestic purposes (other than drinking and cooking), certain types of agricultural applications, stock raising, and industrial consumption. The shorter water resources are, the more important become management manipulations that make sure each drop of water is applied in the best way.

Water quantity. As with any other commodity, for water the target is 'the more the better'. But, having constructed a well, will we always pump from it as much as possible? Not at all. Commonly a steady supply is required. Therefore, water will be pumped at a reasonable rate so the water table (section 4.1) will not drop below the pump inlet and water will flow from the host rocks into the well in a steady rate. Young water is steadily replenished, and with the right pumping rate may be developed into a steady water supply installation. In contrast, pumping old water resembles mining – the amounts are limited. The limitation may be noticed soon or, as for example in the Great Artesian Basin of Australia, the shortage may be felt only after substantial abstraction, but in such cases the shortage causes severe problems due to local overdevelopment based on water. Knowledge of the age of water is thus essential for groundwater management. The topic of water dating is discussed in Chapters 10 and 11.

Hydraulic interconnections. Does pumping in one well lower the output of an adjacent well? How can we predict? How can one test? Everyone knows that water flows down-gradient and, therefore, a well is always suspected to produce water 'on account' of other wells down-gradient (i.e. of lower water tables). But to what distances is this rule effective? And is the way always clear underground for water to flow in all down-gradient directions? The answers to these questions vary from case to case and therefore a direct tracing investigation is always needed. Compositional, age, and temperature similarities may confirm hydraulic interconnections between adjacent springs and wells, whereas significant differences may indicate flow discontinuities (section 6.5).

Mixing of different groundwater types. Investigations reveal an increasing number of case studies in which different water types (e.g. of different composition, temperature, and/or age) intermix underground. The hydrochemist has to explore the situation in each study area.

We are used to regarding groundwater as being homogeneous, which it often is. In such cases a quality analysis represents the whole water body and, similarly, all other parameters measured reflect global properties. But, when fresh water, recharged underground, is mixed with varying quantities of old saline water, the hydrochemist's life becomes tough and interesting. The first task is to find out the properties of the end members of the intermixing waters and their origins (sections 6.6 and 6.7). Then it is possible to recommend proper exploitation policies, balancing the need for the better water type with the demand for large water quantities.

Depth of circulation. Understanding a groundwater system includes knowledge of the depth of circulation. Temperature can supply this information: the temperature of rocks is observed to increase with depth and the same holds true for water kept in rock systems – the deeper water is stored, the warmer it is when it issues in a spring or a well (section 4.7).

Location of recharge. Every bottle of water collected at a spring or a well contains information on the recharge location. Water molecules are separated during precipitation from clouds – molecules with heavier isotopes being rained out more efficiently than molecules with lighter isotopes. As a result, clouds that are blown to high mountains precipitate 'light rain'. Study of the isotopic composition of groundwaters provides clues to the recharge altitude (section 9.8) and, hence, provides constraints on the possible location of recharge.

Sources of pollution. The major topics the hydrochemist has to deal with regarding pollution threats are:

- Early detection of the arrival of pollutants into groundwater in order to provide warning in due time, allowing for protective action.
- Identification of pollutant input sources.

This brings us back to the problem of groundwater intermixings and hydraulic interconnections. It also brings to mind the urgent need for basic data to be collected on non-disturbed systems so that changes caused by contamination may be detected at an early stage.

1.4 Conceptual models

The ultimate goal of a hydrochemical study is the construction of a conceptual model describing the water system in three spatial dimensions with its evolution in time. The conceptual model is mainly qualitative, recognizing recharge areas, modes of water take-up, number and nature of aquifers and aquicludes or karstic regimes, structural and lithological controls, occurrence of mixing and identification of end members, storage capacities, flow directions and velocities, degrees of confinement, depths of circulation, responses to rain and flood events, and sources of pollution. A conceptual model is best summarized in a diagram (section 14.3).

The conceptual model is needed to plan further stages of investigation, to forecast system behaviour and to provide the necessary base for mathematical modelling.

Conceptual models are not only a scientific necessity but are also useful in dealing with clients or financing authorities. A preliminary conceptual model may provide a useful introduction to research proposals, clarifying the suggested working plan and budgetary needs. Conceptual models are useful in summarizing final reports (section 14.3).

1.5 Economic aspects of field hydrochemistry

Basic and applied water research is nearly always conducted on a restricted budget. Traditionally, the larger sums go to the construction of weirs and other discharge measuring devices in surface water studies, and drilling of new wells in groundwater studies. Little money is commonly spent on hydrochemical investigations. The ratio of outcome to costs is, however, in many cases in favour of the hydrochemical studies.

Introduction

New drillings are traditionally conducted to check water occurrences, to establish depth of water tables, and in order to be equipped for pumping new niches of known aquifers. Inclusion of hydrochemical studies as an integral part of the drilling operation may double and triple the amount and quality of information gained, with only a small increase in costs. A number of examples are given below.

Measurements during drilling, providing insight into the vertical distribution of water bodies

Blind drilling is common, aiming at a specific water horizon or aquifer. It is the hydrochemist's task to convince the drilling organization that drilling should be stopped at intervals, water removed so that new water enters the well, and the following measurements carried out:

- Water level.
- Temperature.
- Conductance (indicating salinity).
- A sample should be collected for detailed chemical and isotopical analyses.

This will provide vital information on the vertical distribution of water horizons and their properties (section 7.2). Admittedly, arresting a drilling operation costs money, but the information is indispensible because water layers may otherwise be missed. Savings may be made by conducting hydrochemical measurements and sampling at planned breaks in the drilling operations on weekends and holidays, or unplanned breaks due to mechancial troubles.

Initial database

Whenever a new well is completed, the abstracted aquifer should be studied in detail, including water table and temperature measurements, and complete laboratory analyses of dissolved ions and gases, stable isotopes and age indicators such as tritium and carbon-14. Analyses for suspected pollutants should be carried out, e.g. fertilizers, pesticides and pollutants of local industries. This wealth of data is needed to provide answers to the questions raised in section 1.3.

Pumping changes the underground pressure regime, and fluids begin to move in new directions. Thus, information that was not collected at the very beginning of a well operation is lost, and may not be obtained later on. The hydrologic evolution of flow patterns induced by human activity in exploited water systems can be established by frequent comparisons to the initial situation.

Periodically repeated hydrochemical measurements

There are questions that cannot be answered by the mere placement of new wells. For example:

- Number of water types that are intermixed underground.
- Seasonal variations in water quality and quantity.
- Influence of pumping operations in adjacent wells.

These questions may well be answered by conducting periodically repeated measurements in selected wells and springs. The costs of repeated measurements are fractional as compared to drilling and they provide information not otherwise available, significantly increasing the understanding of the water system in terms needed for its management.

The economical value of hydrochemical studies also has wider aspects; they may reduce the number of new drillings needed, and they may place them at better locations. The price of hydrochemical measurements is small compared to drilling expenses but, never the less, they have to be minimized by careful planning and proper selection of parameters measured at each stage of the study, and proper selection of water sampling frequencies. These topics are discussed in Chapter 7.

2 BASIC HYDROLOGICAL CONCEPTS

2.1 The aerated intake zone

A typical well passes through two zones: an upper zone, usually of soil-covered rock, that contains air and water in pores and fissures, and a lower zone, often of rock, that contains only water in pores and fissures. This separation into two parts is observed everywhere, and the two zones have been named the *aerated* (air-containing) *zone* and the *saturated* (water-saturated) *zone*. The top of the saturated zone (where water is hit in a well) is called the *water table* (Fig. 2.1). The water table is described either as the depth from the surface, or as the altitude above sea level (section 2.2).

The aerated zone is also called the vadose zone (from Latin *vadosus*, shallow). Its thickness varies from zero (swamps) to several hundred metres (in regions of elevated and rugged topography and arid climates). Common thicknesses are 5–25 m. In most cases the aerated zone has two parts: soil at the top and a rocky section beneath.

The soil, formed by weathering and biogenic processes, varies in thickness from zero (bare rock surface) to a few metres, reaching maximum thickness in alluvial valleys. The soil zone is characterized by the plants that grow in it, their roots penetrating to 10–40 cm, but certain plants, mainly trees, have roots that grow as deep as 20–30 m. Roots may penetrate the lower rocky part of the aerated zone, occasionally reaching the water table. The plants respire and liberate CO_2 into the soil pores. Bacteria decompose plant remains in the soil, adding to the CO_2 of the soil air. This biogenic CO_2 dissolves in infilatrating water, producing carbonic acid, which slowly but continually dissolves minerals and rock components. This process creates soil and provides recharging water with dissolved salts (sections 6.8 and 6.9).

Intake of water in the aerated zone is either by infiltration into the soil cover or, in bare rock surfaces, by infiltration into intergranular pores (as in sandstone), fissures and joints (as in igenous rocks or quartzites), or dissolution conduits and cavities (limestone, dolomite, gypsum, rock salt). Only pores and fissures that are interconnected, or communicate, are effective to infiltration.

Water infiltration in a sandy soil may be described as water movement in a sponge, or homogeneous granular medium. Water descends in this setting in downward moving fronts, or in a piston-like flow (Fig. 2.2). In such a mode of flow a layering exists, the deeper water layers being older than the shallow ones. Measurements reveal that such piston flows proceed at velocities of a few centimetres to a few metres a year. This velocity depends on the porosity of the soil, the degree of pore interconnection, amount and distribution of local precipitation, and drainage conditions. In contrast to the sponge-type flow, recharge into the ground is often rapid, through conduits and zones of preferred transmissivity. Examples of such conduits are open dessication cracks and old dessication cracks that have filled up with coarse material, living and decayed roots, animal holes, or bioturbations, and gravel-packed channels of buried stream beds (Fig. 2.3). Karstic terrains (section 2.7) provide extreme examples of recharge via conduits. In most cases water descends through the aerated zone in a combined mode, i.e. through intergranular pores and also through conduits that cross the granular medium (Fig. 2.4).

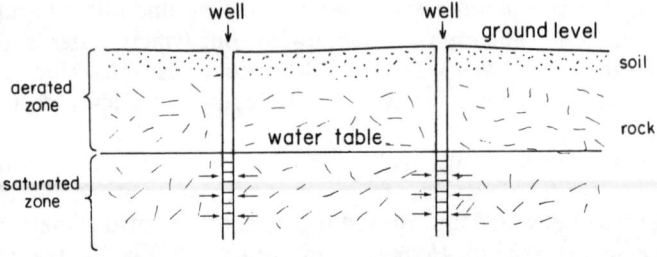

Fig. 2.1 A schematic section through typical wells: at the upper part the soil and rocks contain air and water in pores and fissures, forming the aerated zone. Below occur rocks with only water in their pores and fissures, forming the saturated zone. The top of the saturated zone is the water table, recognizable in wells as the depth at which water is encountered.

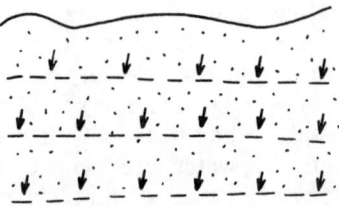

Fig. 2.2 Piston flow of water infiltrating into a homogeneous granular soil ('sponge'-type flow). A layered structure is formed, lower water layers being older than shallow ones. Penetration velocities are observed to range from a few centimetres to a few metres per year.

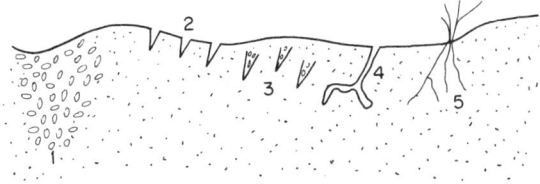

Fig. 2.3 Enhanced recharge through parts of high conductance in soil, alluvium and non-consolidated rocks: (1) gravel in buried stream beds; (2) open desiccation cracks; (3) old desiccation cracks filled with coarse material; (4) animal holes; (5) plant roots.

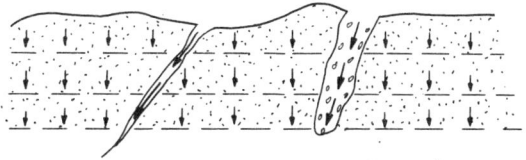

Fig. 2.4 Infiltration of recharge water in a combined mode: part of the water moves slowly through the granular (mainly soil) portion of the aerated zone, and part moves fast through open conduits or zones of high transmission.

2.2 The saturated zone and the water table

Infiltrating water settles by gravity, reaching a depth at which all voids in the ground are filled with water. This portion of the ground is called the saturated zone. The water in this zone forms a continuous medium, interwoven with the solid matter system.

The water in the saturated zone settles with a smooth horizontal upper face, called the water face, or water table. By analogy, water poured into a vessel settles with a smooth horizontal upper face – a miniature water table. The depth of the water table, or the water level, is expressed either in relation to the land surface, or as the altitude above sea level. The depth of water tables varies from zero, in marshy areas, to several hundreds of metres, in dry mountainous terrains. A depth of a few tens of metres is common.

Measurements of the water table are of prime interest for management purposes and for the establishment of flow directions. Study of water table fluctuations is most valuable: rapid response to rain, flood events, or snow melt indicates fast recharge via conduits. Delayed response indicates slow infiltration through a porous medium. Changes in observed water table, caused by pumping in adjacent wells, establish the existence of hydraulic interconnections between the respective wells (sections 4.1 and 4.2).

The slope of the water table is defined by three major factors: the base of the drainage system (sea, lake), the topography of the land surface and the occurrence of impermeable rock beds.

The saturated zone usually contains strata of varying permeabilities, forming aquifers and aquicludes, described in the following sections.

2.3 Aquifers

Rock beds at the saturated zone that host flowing groundwater are called aquifers (from Latin: *aqua*, water; *ferre*, to bear, or carry). Aquifer rocks contain the water in voids – pores and fissures. The size and number of voids, and the degree of interconnection between these pores and fissures, define the qualities of the aquifers. The same properties have been shown above to define infiltration efficiency and capacity of intake of recharge water. These properties are discussed below for different rocks.

Conglomerates. These rock deposits are made up by rock pebbles, cemented to different degrees. If poorly cemented, conglomerates are extremely efficient water-intake systems; they make aquifers with large storage volumes, and water flows relatively fast through them (provided gradients are steep enough and drainage conditions are good). Often conglomerates are cemented, by calcite, iron oxides, or silica. Cementation reduces the water-carrying capacity of an aquifer, and in extreme cases conglomerates are so tightly cemented that they act as aquicludes.

Sandstone is in most cases composed of quartz grains with pores constituting 10–50% of the rock volume. The pores are interconnected, providing high infiltration efficiencies and providing sandstone aquifers with high storage capacities and rapid water throughflow. Cementation reduces the pore space, but cemented rocks tend to develop mechanically formed fissures which let water through.

Limestone and dolomite. These carbonate rocks are often well crystallized and can be poor in interconnected pores. However, limestone and dolomite tend to fracture under tectonic stress and through such fractures groundwater can move. Water can have two completely different effects on the fractures: either the water fills up the fissures by precipitation of carbonates or, under high flow conditions, the water dissolves the rock and causes further opening. Dissolution of limestone and dolomite may thus create surface and subsurface conduits that enhance intake of recharge water and form high-conducting aquifers with large water storage capacities. Karstic systems are the result of extreme dissolution activity in hard carbonate rocks (section 2.7).

Igneous rocks. Granite and other intrusive igneous rocks have intrinsically well-crystallized structures with no empty pores. However, being rigid they

Basic Hydrological Concepts

tend to fracture and may have a limited degree of infiltration capacity and aquifer conductivity. Weathering in humid climates may result in extensive formation of soil but little dissolution opening of fractures. Thus, granitic rocks may form very poor to medium quality aquifers.

Basalt and other extrusive lava rocks are similar to granitic rocks in having no pores, but tending to fracture. In addition, lava rocks often occur in bedded structures, alternating with paleosoils and other types of conducting materials, e.g. rough lava flow surfaces.

Thus, basaltic terrains may have medium lateral flow condutivities. Weathering of basaltic rocks and paleosoils produces clay minerals (section 6.8) that tend to clog fractures. Tuff (a rock formed by fragmental volcanic ejecta) is porous when formed, but is readily weathered into clay-rich impervious soils.

Metamorphic rocks. Quartzite, schist, gneiss, and other metamorphic rocks and metasediments vary in their hydrological properties, but in most cases are fractured and transmit water. In rainy zones weathering may produce clays that occasionally clog the fractures.

The rocks discussed in the present section have intrinsically medium to good infiltration efficiency and aquifer conductance and storage capacities. These features may be developed to different degrees, as a result of secondary processes, such as tectonic fracturing and chemical dissolution opening on the one hand, and clogging by sedimentation and weathering products on the other hand. Thus, close study of types of rocks and of accompanying features are essential parts of hydrological investigations.

2.4 Aquicludes

Rock strata that prevent passage of groundwater are called aquicludes (from Latin: *aqua*, water; *claudere*, to close). Aquicludes are important components of groundwater systems, because they seal the aquifers and prevent water from infiltrating to great depths. Aquicludes are essential to the formation of springs and of shallow accessible aquifers. Aquiclude rocks have low water conductivity, caused by lack of interconnected voids or conduits. A number of common aquiclude-forming rocks are discussed below.

Clay. Clay minerals have several characteristics that make them good aquicludes:

- When consolidated, they have effectively no interconnected pores.
- They swell, closing desiccation fractures or tectonic fractures.
- They are plastic - another property that makes clays effective sealing agents that fill up potenial openings.

The effectiveness of clay aquicludes depends on the type of clay, montmorillonite being perhaps the best. The thickness of the clay aquiclude is of prime importance – occasionally a clay bed 1 m thick may be a significant barrier to underground water movement, but water may slowly leak through. The thicker the aquiclude, the higher is its sealing efficiency. Clay aquicludes with a thickness of several hundred metres are known.

Shales. Shales are derivatives of clays, formed in slow diagenetic processes. Shales have little or negligible swelling capacity, and have medium plasticity. They may form effective aquicludes at a thickness of at least a few metres to tens of metres.

Igneous rocks. Large igneous bodies, such as stocks or thick sills, often act as aquicludes because they lack interconnected fractures or dissolution conduits.

Dykes and sills may act as aquicludes, if they are either very fresh and not fractured, or weathered into clay-rich rocks.

2.5 Phreatic, or free surface, aquifers

Phreatic aquifers have free communication with the aerated zone. The synonym 'free surface aquifer' relates to the free communication between the aquifer and the vadose zone. An example is shown in Fig. 2.5. The term phreatic originates from the Greek word for a well.

Phreatic aquifers are the most exploited type of aquifer, and most of the hydrochemist's work is performed on them. Phreatic aquifers are the collectors of infiltrating recharge water, and this process is well reflected in the chemical, physical, and isotopic parameters of the aquifer's water: water table fluctuations indicate the nature of recharge intake (sections 4.1 and 4.2), water table gradients indicate flow directions (section 4.3), water temperature indicates nature of infiltration (section 4.7) and depth of circulation (section 4.7), and tritium and radiocarbon concentrations reflect water ages and modes and velocity of flow (sections 10.7 and 10.8).

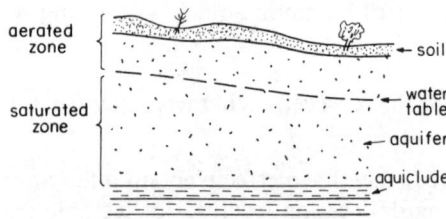

Fig. 2.5 Basic components of a phreatic groundwater system: intake outcrops, an aerated zone, the water table, the saturated zone that constitutes a water-bearing aquifer, and impermeable rock beds of the aquiclude that seal the aquifer at its base.

2.6 Confined aquifers and artesian flow

Confined aquifers are water-bearing strata that are sealed at the top and the bottom by aquiclude rocks of low permeability (Fig. 2.6). Confined aquifers are commonly formed in folded terrains (section 3.4) and have a phreatic section, where the aquifer rock beds are exposed to recharge infiltration, and a confined section where the aquifer rock beds are isolated from the landscape surface by an aquiclude (Fig. 2.6).

The water in the saturated zone of the phreatic section of a confined system exerts a hydrostatic pressure that causes water to ascend in wells. In fact, a confined aquifer can often be identified by the observation that water ascends in a borehole to a level higher than the level at which the water was first struck. In extreme cases the water ascends to the surface, constituting an artesian well. This phenomenon of water ascending in a well and flowing 'by itself' was first described in 1750 in the area of Artois, a northern province of France. The term *artesian* is applied to self-flowing wells and to aquifers supporting such wells. The term *semi-artesian* describes wells in which water ascends above the depth at which the water was struck but does not reach the surface.

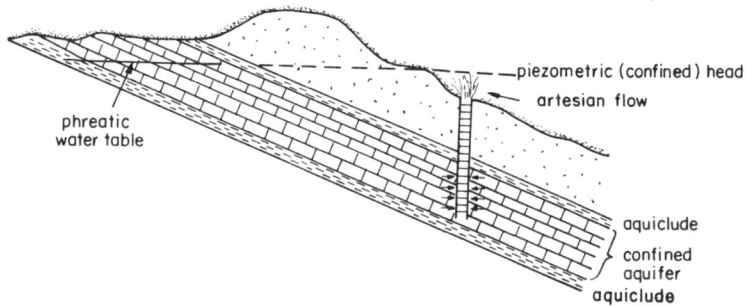

Fig. 2.6 Components of a confined aquifer: tilted, or folded, water-bearing rock strata, sealed at the top and the base by aquicludes. Each active confined system also has a phreatic section, at outcrops of the aquifer rocks. The level of the water table in the phreatic section defines the piezometric head in the confined section. Water ascends in boreholes drilled into confined aquifers. Water reaches the surface in artesian flow in boreholes that are drilled at altitudes lower than the piezometric head.

The level water reaches in an artesian well reflects its pressure, called the *piezometric*, or *confined, water head* (Fig. 2.6). In boreholes drilled at topographic altitudes that are lower than the piezometric head, water will reach the surface in a jet (or wellhead pressure) proportional to the difference between the altitude of the wellhead and the piezometric head. The piezometric head is slightly lower than the water level in the relevant phreatic section of the system, due to the flow resistance of the aquifer. Confined aquifers often underly a phreatic aquifer, as shown in Fig. 2.7.

The nature of such groundwater systems may be revealed by data measured in boreholds and wells: the water levels in wells 1 and 2 of Fig. 2.7 did not rise after the water was encountered, and both wells reached a phreatic aquifer. Well 3 is artesian, and the drillers' account should include the depth in which the water was struck and the depth and nature of the aquiclude. The hydraulic interconnection between well 1 and well 3 may be established

- By checking their water heads.
- By the resemblance of the relevant aquifer rocks.
- By agreement of the measured dips at the outcrop and the depth at which the water was encountered at well 3.

Flow of groundwater in confined aquifers is determined by the water head gradient (section 2.8), and also by the degree to which the system is drained. In extreme cases water may be entrapped in confined aquifers. The topic of water movement in confined aquifers is best studied by means of the chemical and isotopic parameters (section 6.5).

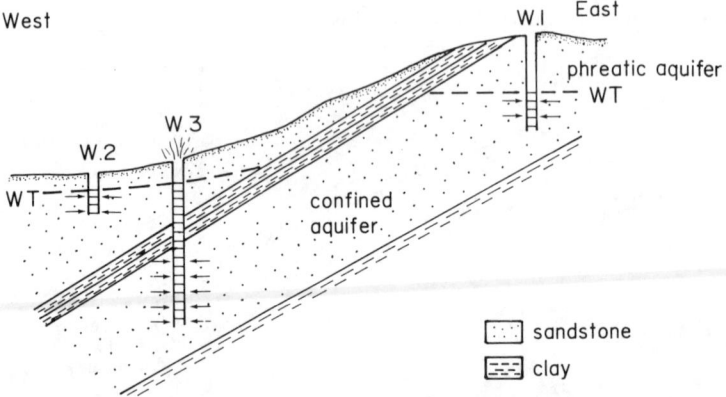

Fig. 2.7 A confined aquifer underlying a phreatic aquifer. The nature of such a system may be established by parameters measured at boreholes and wells (see text).

2.7 Karstic systems

Hard calcereous rocks, i.e. limestone and dolomite, develop specific features noticeable on the land surface: dissolution fissures, cavities, caves, and sinkholes in which runoff water disappears. On the slopes of such terrains occur springs with high discharge that often undergo a marked seasonal cycle. In such terrains intake of recharge water and underground flow are in open conduits (sections 2.1 and 2.2). A classical area of such conduit-controlled water systems has been described in the Karst Mountains of Yugoslavia, and from there originated the term *karstic*, pertaining to groundwater flow in dissolution conduits. Some basic features of karstic systems are depicted in Fig. 2.8. Sinkholes are inlets to conduits in which runoff water can move rapidly into deeper parts of the system. Karstic springs occur at the base of local drainage basins, commonly major river beds. Abandoned spring outlets remain 'hanging' above new drainage bases, forming caves which can sometimes be rather deep. Dissolution conduits are formed by the chemical interaction of CO_2 charged water with calcereous rocks (section 6.8).

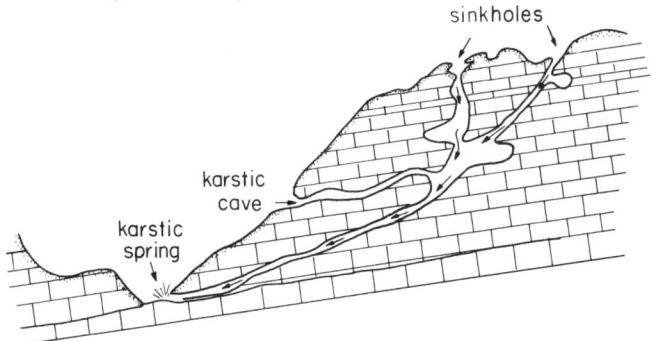

Fig. 2.8 Components of a karstic system: sinkholes and fissures acting as inlets for rapid intake of runoff water; dissolution channels; water discharging in springs at local bases of drainage, e.g. river beds; caves exposed at former dissolution channels, formed at former (higher) drainage bases.

Karstic systems are often formed in 'generations'. For example, during the Pleistocene, when the oceans had a lower water level, low-level karstic systems were formed in coastal areas. With the rise of sea level, higher karstic systems evolved.

The karstic nature of recharge intake zones is often recognizable from specific geomorphic features, but this is not always the case: karstic systems are occasionally concealed and karstic outlets may occur tens or even hundreds of kilometres from the intake areas. Furthermore, paleokarstic conduits may be in operation in regions which at present have a semi-arid climate.

Thus, indicators for karstic flow are needed. In small-scale studies, direct tracing of karstic underground interconnections has been successfully carried out with dyes, spores, and radioactive tracers. The limitations are:

- The dependence on sinkholes as tracer inlets.
- The logistics of tracing experiments that limits them to distances from hundreds of metres to, at most, a few kilometres.

Hydrological and hydrochemical indicators for karstic flow are:

- Rapid flow, revealed by very young groundwater ages (months to a few years).
- Water temperatures that maintain the intake temperatures (e.g. snow melt temperatures).
- Significant seasonal discharge fluctuations.

- Light stable isotope components, reflecting composition of high-altitude precipitation and indicating lack of evaporation.
- Chemical composition reflecting contact with limestone and/or dolomite (section 6.8).

2.8 Velocity of groundwater flow in the saturated zone

Water is *stored* in rocks mainly in pores. The *effective porosity* of a rock is the volume percentage of the rock that may contain water in pores. The values (Table 2.1) are high for non-consolidated granular rocks e.g. soils, clay, silt, and gravels. The porosity is low for crystallized rocks such as limestone, dolomite, and most igneous rocks. *Movement* of water through rocks is through interconnected pores, fissures, and conduits. *Permeability* is a measure for the ease of flow through rocks. (Hydraulic conductivity is another term.) The permeability relates to pores and voids that are interconnected, and it differs from porosity, which is the total pore volume. Clay has a high content of water, stored in the molecular lattice, but these water sites have practically no interconnections. Thus, in spite of its high capacity for containing water, clay has a low permeability, making it a most efficient aquiclude.

In contrast, sandstone (non-consolidated) has a high permeability, providing excellent aquifer properties. The permeability coefficients of common unconsolidated sediments vary, in order:

gravel > sand > silt > clay.

Table 2.1 Rock porosities, volume per cent

Soils	50-60%	Fine sand	30-35%
Clay	40-55%	Pebbles	30-40%
Silt	40-50%	Sandstone	10-20%
Coarse sand	35-40%	Limestone	1-10%

Darcy's law. The velocity, V, by which water moves between two points in an aquifer is proportional to:

- The hydraulic gradient, i.e. the height difference, dh, divided by the distance (i.e. length of flow path), dl (Fig. 2.9), and
- The coefficient of permeability K.

This relation was studied by Darcy in a small scale experiment, but it is applied on a large scale to flow in groundwater systems:

$$V = K \frac{dh}{dl} \quad \text{(Darcy's Law)}$$

Basic Hydrological Concepts

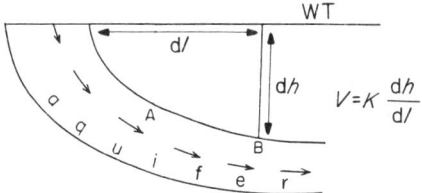

Fig. 2.9 Darcy's law, applied to a groundwater system: the velocity V, by which water flows in the aquifer rocks, is proportional to the coefficient of permeability, K, and the gradient, dh/dl. This is a greatly simplified approach, and a number of assumptions have to be checked with hydrochemical methods (see text).

The coefficient of permeability, K, is empirically deduced from pumping tests and laboratory measurements on rock cores. Its value varies laterally and vertically, and the average value is often disputed by orders of magnitude. The difference between the water levels is obtained by water table measurements in wells (section 4.1); and the distance is measured, or read from a map.

Several assumptions underlie the application of Darcy's law to groundwater systems:

- The system is homogeneous, and properly described by the applied permeability coefficient.
- The system is hydraulically continuous, with no barriers to the suggested flow, and well represented by the accessible wells.
- The system has free drainage, a property that is often restricted, or lacking, mainly in deep confined groundwater systems.

A direct check of these assumptions in the field is needed, a topic dealt with in sections 4.4, 4.8, 6.5, and 11.9. Water velocities calculated on hydraulic grounds should be checked by other methods: among these, isotopic dating methods are of prime importance (sections 10.4, 10.7, 11.8 and 12.9).

The topic of groundwater flow and related concepts has been briefly discussed in a simplified manner, adequate for the hydrochemical and isotopic approaches discussed in the present book. A more thorough treatment, from a water engineering point of view, can be found in Todd's textbook *Groundwater Hydrology* (1980) or de Marsily's book *Quantitative Hydrology* (1986).

2.9 Modes of groundwater flow in the saturated zone

Groundwater flows in the saturated zone through interconnected pores as well as through interconnected conduits (fissures and dissolution cavities). These types of flow, and combinations of them, are similar to the modes of flow described in section 2.1 for the aerated zone (Figs. 2.1 and 2.2).

Rock fissures are formed mainly in tectonic processes. These fissures may be opened by dilation, or closed by compression or plastic flow. Marl is a good example of a plastic rock lacking open fissures. On the other hand, many limestones and most dolomites are hard and rigid, maintaining open fissures. Water moving in fissures enlarges them by dissolution or, occasionally, fills them up by the precipitation of mineral veins. Karstic conduits are the result of extreme dissolution.

Geologists have limited means of establishing the occurrence of dissolution conduits, as information from drilling cores is scarce.

Hydrochemistry provides the most suitable means of distinguishing conduit-dominated flow from pore-dominated flow. In the first case, discharge and water ages vary significantly in adjacent wells, whereas in the second case discharge and water age are rather constant for a given section of an aquifer. In certain springs a combination of a steady base flow and a seasonly varying water component is observed, the former resulting from pore-dominated flows and the latter being recharged via karstic conduits.

More on the hydrochemical approach to the study of flow in fissures is to be found in a special publication of the international Atomic Energy Agency: *Isotope Techniques in the Study of the Hydrology of Fractured and Fissured Rocks* (1986).

2.10 Evapotranspiration

Only part of the water infiltrating into the ground keeps moving downward into the saturated groundwater zone. An important part of infiltrated water is transferred back into the atmosphere. Two major mechanisms are involved: evaporation and transpiration, together named evapotranspiration.

Evaporation is a physical process, caused by heat energy input, providing water molecules with kinetic energy that transfers them from an inter-pore liquid phase into a vapour phase. Evaporation depends on local temperature, humidity, wind, and other atmospheric parameters, along with soil properties. A maximum value is provided by the extent of evaporation from an open water body, and this varies from place to place in the range of 1–2 m per year. Taken at face value, this would imply that no effective recharge occurs in areas that have an average annual precipitation rate lower than the potential evaporation rate measured in an open water body. This is not the case, and effective recharge is known from regions that have as little as 70 mm average annual rainfall. This indicates that water infiltrated into the ground is partially protected from evaporation. In fact, evaporation occurs mainly from the uppermost soil surface and, when this is dry, further evaporation depends on capillary ascent of water, which is a slow and only partially efficient process. It seems that water which has infiltrated to a depth of 1 m is almost 'safe' from capilary ascent and evaporation. From this discussion it becomes clear that the relative contribution of precipitation to recharge is proportional to the amount of rain falling in each rain event. Intensive rain events push water deep into the ground, contributing to recharge, whereas sporadic rains only wet the soil and the water is then lost by subsequent evaporation.

The mode of rainfall and degree of evaporation influence groundwater composition. High degrees of evaporation result in enrichment of the heavier stable isotopes of hydrogen and oxygen (sections 9.3 and 9.6), and in higher concentrations of dissolved salts.

Transpiration is the process by which plants lose water, mainly from the surfaces of leaves. Plants act as pumps, their roots extracting water from soil and the leaves transpiring it into the atmosphere. Thus, the depth of the transpiration effect on soil moisture is defined by the depth of the root system. The latter is most intensive down to 40 cm and drops to almost nil at depths greater than 2 m. Exceptionally, roots of certain plants, mainly trees, are occasionally observed to penetrate to 20 or even 30 m.

2.11 Recharge

The term recharge relates to the water added to the groundwater systems, i.e. to the saturated zone. Thus, *recharge* is the balance between the amount of water that infiltrates into the ground and the evapotranspiration losses. The prefix emphasizes the repeated action of the water cycle, of which recharge is a segment.

Location of recharge areas is of prime importance in modern hydrochemistry, as such areas have to be protected in order to preserve groundwater quality. Urbanization, agriculture, and almost any other kind of human activity, may spoil the quality of recharge water. Limitations on the use of fertilizers and pesticides in recharge areas are often enforced, as well as limitations on mining and other earth-moving activities. As such limitations have financial impacts, an increasing number of cases are taken to court and hydrochemists may be asked to deliver expert opinions involving the identification of areas of active recharge.

The following are a few features characterizing recharge regions:

- Depth of water table is at least a few metres below surface.
- Water table contours show a local high.
- Significant seasonal water table variations are noticed.
- Local groundwater temperatures are equal to, or several degrees colder than, average ambient annual temperature.
- Effective water ages are very recent (a few months to a few years).
- Salt content is low in most cases in non-polluted areas (less than about 800 mg/l total dissolved ions).
- Surface is covered by sand, permeable soil, or outcrops of permeable rocks.

2.12 Discharge

The term *discharge* relates to the emergence of groundwater to the surface as springs, water feeding swamps and lakes, and water pumped from wells. Discharge is the output of groundwater. Rate of discharge is measured in units of volume per time, e.g. m^3/h (section 4.10).

Discharge areas warrant identification in relation to drainage operations, water exploration and water quality preservation.

Several features characterize discharge areas:

- Water table levels mark, in most cases, a low in the water table contour map.
- Water temperatures are a few degrees above average annual ambient temperature, reflecting heating by the thermal gradient while circulating at depth (section 4.8).
- Waters are either fresh, or saline (if they have passed through salinizing rocks, section 3.1).

3 GEOLOGICAL DATA

3.1 Lithology and its bearing on water composition

The composition of rocks and soils has a direct bearing on the quality of water. Hydrochemically, a rough classification into three rock groups seems practical:

- Rocks in which fresh groundwater is common, i.e. rocks that contribute extremely little salts to the water.
- Carbonate rocks that contribute dissolved matter but maintain good potable quality.
- Rocks that enrich the water with significant amounts of dissolved salts, often making them non-potable.

The lithological parameter is only one among several parameters that control groundwater quality. Other factors may be evaporation at the surface prior to infiltration transpiration, wash-down of sea spray, and reducing conditions in the aquifer, connected to H_2S production. Water moves underground, and its salt or mineral content is determined by all the soil and rock types it passes through. Thus, occasionally, saline water may be encountered in rocks that do not themselves contribute soluble salts.

Rocks that preserve high quality of water. Pure sandstone, rich in quartz grains and with little or no cement, may host very fresh water with very little addition to the salts brought in by recharging rain (Table 3.1).

Other rocks hosting fresh water are quartzite (sandstone cemented by silica), basalt, granite, and other igneous rocks (Table 3.1).

The total dissolved salts in water encountered in these rocks is often in the range of 300–500 mg/l.

Carbonate rocks commonly host water with 500 to 800 mg/l total dissolved salts, mainly $Ca(HCO_3)_2$ in the case of limestone and $(Ca, Mg)(HCO_3)_2$ in the case of dolomites. The relevant chemical water-rock interactions are discussed in section 6.8.

Water encountered in limestone and dolomite is generally tasty and of high quality.

Table 3.1 Lithological imprints on groundwater composition

Rock	Groundwater composition
Sandstone	Low salinity (300–500 mg/l); HCO_3^- major anion, Na^+, Ca^{2+}, Mg^{2+} in similar amounts; good taste.
Limestone	Low salinity (500–800 mg/l); HCO_3^- major anion, Ca^{2+} dominant cation; good taste.
Dolomite	Low salinity (500–800 mg/l); HCO_3^- major anion; Mg^{2+} equals Ca^{2+}; good taste.
Granite	Very low salinity (300 mg/l); HCO_3^- major anion, Ca^{2+} and Na^+ major cations; very good taste.
Basalt	Low salinity (400 mg/l); HCO_3^- major anion; Na^+, Ca^{2+}, Mg^{2+} equally important; good taste.
Schist	Low salinity (300 mg/l); HCO_3^- major anion; Ca^{2+} and Na^+ major cations; good taste.
Marl	Medium salinity (1200 mg/l); HCO_3^- and Cl^- major anions, Na^+ and Ca^{2+} major cations; poor taste but potable.
Clay and shale	Often containing rock salt and gypsum. High salinity (900–2000 mg/l); Cl^- dominant anion, followed by SO_4^{2-}; Na^+ major cation; poor taste, occasionally non-potable.
Gypsum	High salinity (2000–4000 mg/l); SO_4^{2-} dominant anion; Ca^{2+} dominant cation, followed by Mg^{2+} or Na^+; bitter, non-potable.

Evaporites – gypsum and salt rock. Gypsum ($CaSO_4.2H_2O$) is formed mainly in lagoons (semi-closed shallow water bodies, mainly along oceanic coasts) that are subjected to high degrees of evaporation. Sea water evaporated to ⅓ of its original volume precipitates gypsum (or anhydrite, the water-poor variety of gypsum). Groundwater in contact with gypsum rock dissolves it, along with accompanying salts, reaching 1000 to 2600 mg/l of SO_4 along with Cl, Ca, Mg and Na. The total dissolved salt content can reach 2000–4000 mg/l (Table 3.1). The taste of such water is bad, mainly bitter. Water with up to 600 mg/l SO_4 may be consumed in cases of emergency, but higher concentrations will cause thirst rather than quench it. Stock may, however, drink such water. Rock salt (NaCl) is formed by even higher degrees of evaporation, also in lagoons and closed lakes. Sea water is saturated in NaCl when evaporated to ¹⁄₁₀ of its original volume. Water coming in contact with rock salt dissolves it, becoming enriched in NaCl. Water in contact with rock salt is non-potable, with salinity up to that of sea water (brackish water) or saltier than sea water (brine).

Clay and shale hosting gypsum and rock salt. Clay and shale are hydro-aluminium silicates that by themselves practically do not add salts to water that comes in contact with them. However, clay and shale often contain veins and nodules of gypsum, pyrite, and rock salt. Clay and shale are impermeable and form aquicludes rather than aquifers (sections 2.3 and

2.4), but due to the high solubility of gypsum and especially rock salt, groundwater in contact with clay and shale, at the base of aquifers, often become saline and poor in quality.

Lithological considerations in well location and design – the water quality aspect. Selection of well sites is made on the basis of several aspects, water quality being a major one. Thus, a thorough knowledge of the lithological section of candidate well sites is of prime importance. Abstraction from massive gypsum and salt rocks should be avoided in any case. Aquicludes of salinizing rocks, such as clay and shale, have to be avoided, either by terminating the wells or their perforated sections above the clay or shale layer, or by sealing the well at the aquiclude section, avoiding entrance of poor-quality water (Fig. 3.1).

Additional information on water composition as a function of aquifer rock composition is found in the books of Davis and De Wiest (1966), Matthess (1982), Drever (1982), or Eriksson (1985).

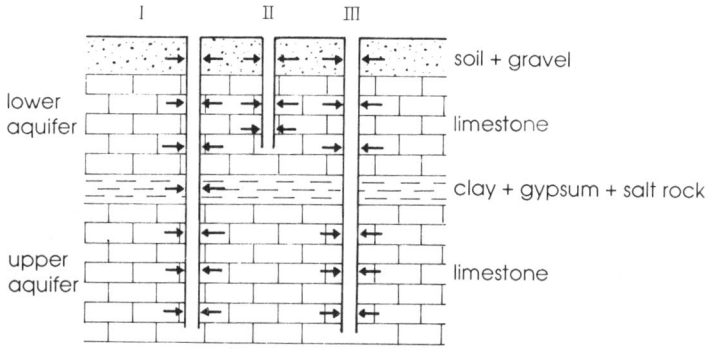

Fig. 3.1 Two aquifers of non-saliferous rocks, separated by an aquiclude of clay with gypsum and salt rock. Well I – fully perforated and producing saline water; well II – stopped at safe distance above the clay, abstracting good water from the upper aquifer alone; well III – sealed for several metres above and below the clay bed, producing good water from both aquifers.

3.2 Properties of geological materials and their bearing on recharge and groundwater storage

Infiltration of recharge water occurs through interconnected pores and open fissures, or combinations of the two (Fig. 3.2). In contrast, rocks rich in pores that are too small or isolated from each other, non-porous rocks, and rocks with no open fissures, are non-efficient for infiltration and recharge.

The same principles hold true for water movement and water storage in aquifers. Rocks with open interconnected pores and open interconnected fissures are best for recharge and storage.

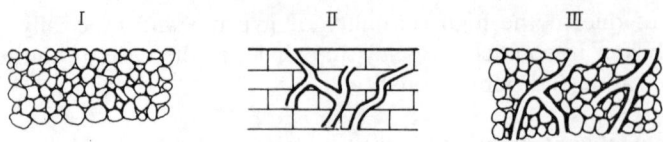

Fig. 3.2 Modes of water infiltration. I – porous texture with no cement between the grains, high infiltration and water storage capacity; II – non-porous texture with open interconnected fissures, good infiltration and storage capacity; III – partially porous texture with interconnected fissures, infiltration and water storage are possible in two modes: through non-cemented interconnected pores and through fissures.

Materials with open interconnected pores include all granular rocks. The most common types are sand and sandstone. Their recharge and storage quality increase with the grain size. Cementation (carbonate, silica, iron oxide) reduces the pore volume and, in extreme cases, may result in an impermeable sandstone (although cemented sandstone tends to fissure).

Conglomerate (a rock made up of pebbles) is an extreme case of a granular rock and, if non-cemented, has excellent recharge and storage properties.

Materials with open interconnected fissures. All brittle rocks tend to be fractured to different degrees. The joints and fissures may in certain cases be somewhat enlarged by dissolution, but they may also be clogged with clays produced by weathering. Thus, it is not enough to know the rocks that constitute the terrain in studied areas, it is also necessary to know their degree of fracturing and of weathering.

An extreme case of open interconnected conduits is manifested by limestone and dolomite in karstic terrains, as described in section 2.7.

Materials with poorly connected pores include clay and shale. They have a large amount of minute pores, totalling up to 55% of the rock volume. Yet these pores are poorly interconnected, resulting in low permeability. Clay and shale significantly slow down infiltration and serve as aquicludes.

Materials with isolated fissures. Metamorphic rocks are rich in open foliation, fractures, and fissures, but these are isolated or poorly interconnected. Such rocks make poor recharge terrains and poor aquifers. They are occasionally fractured by tectonic processes, improving their infiltration and storage properties.

Studying rocks for their conductivity, or recharge and storage capacities. Rock properties should be examined at as many outcrops as possible – looking for friability, cementation, degree and nature of fractures, dissolution conduits, and animal burrowing (Figs. 2.3, 2.4, 2.8, and 3.2). Similar observations may be conducted on drill cores, but these are

Geological Data 27

expensive and their record is limited in size. Laboratory tests on cores provide semi-quantitative data on the nature of rock pores, mini-fissures and conductivities.

Another way to determine the relevant rock properties is by examining the properties of groundwater samples in springs and wells. Low water ages (i.e. young water), and seasonal variations of discharge and temperature, indicate rapid passage through high-conducting rocks, whereas high water ages (i.e. old water), and constant discharge and temperature, indicate retardation and low conductivity, at least along part of the water system. These, and other relevant water tracing approaches, are the topics of the following chapters.

3.3 Layering and its effect on groundwater flow

Many rock types have a layered structure, individual rock layers varying in thickness from a few centimetres to tens of metres. The layered rocks include the marine sediments, most continental sediments, lava flows, and volcanic ejecta and intrusive sills. The hydraulic properties vary from one rock layer to the other, often resulting in abrupt changes along the vertical axis. In terms of the permeability coefficient, K, the lateral coefficient K_x may significantly differ from the vertical coefficient K_z. The alternation of aquifers and aquicludes results from the layered structure of different rocks, and the occurrence of springs is often controlled by the layering of rocks.

Fissures may be restricted to individual rock layers, or cross several rock beds. In the later case water flow is improved, mainly in the vertical direction.

3.4 Folded structures and their bearing on flow direction and confinement

Water flows from high to low points, and prefers paths of least resistance. Thus, in regions built of bedded rocks, tilting may be a prime factor, determining flow direction (besides topographic gradients) as shown in Fig. 3.3. Among folded structures, synclines are important as they often gather water from the neighbouring anticlinal structures (Fig. 3.4). Hence, understanding the tectonic regime of a well target area is essential.

Flow directions deduced from tectonic and topographic arguments must always be checked by other methods, as outlined in the following chapters.

Confined aquifers (section 2.6) are rare in tectonically undisturbed regions with horizontal rock beds (Fig. 3.5). Tilting of the aquifer and aquicludes 'sandwich' makes room for the formation of confined aquifers. It provides each case with a recharge outcrop section, forming a phreatic aquifer (section 2.5) and a confined section, fed by the former (Fig. 3.6).

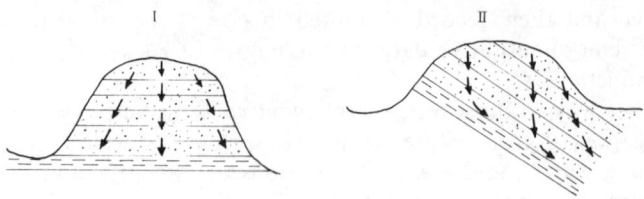

Fig. 3.3 Factors controlling the flow direction of groundwater: I - porous rock - water direction is determined by topographic gradients alone; II - a tilted impervious rock bed deviates direction of water flow.

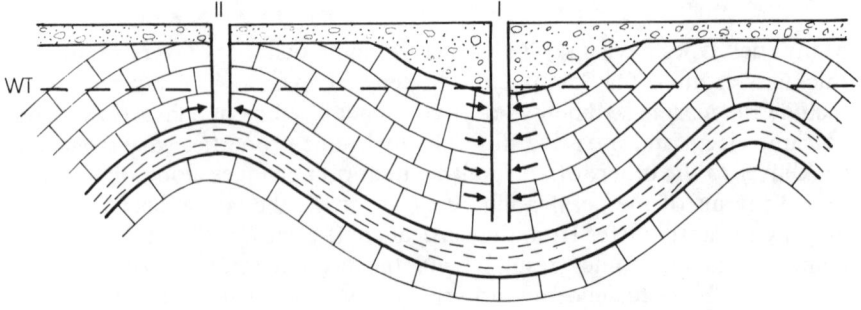

Fig. 3.4 Groundwater accumulating in a syncline. Well I, located at the syncline, has a significantly higher production than well II, located in the adjacent anticline. A drop in the water table, due to pumping, will cut well II dry in a short time. On the other hand, fracturing in the crest of an anticline occasionally makes these locations the better place for water supply.

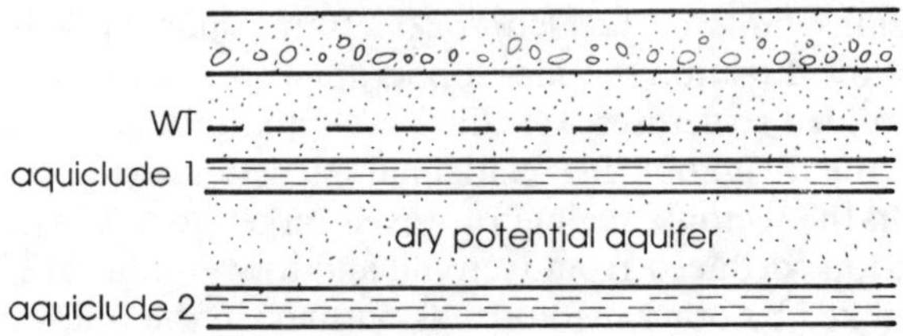

Fig. 3.5 Horizontal rock beds may allow for the formation of only one phreatic aquifer, the first aquiclude preventing the water from reaching lower potential aquifers.

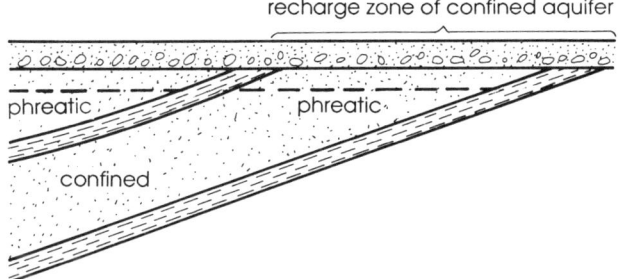

Fig. 3.6 Tectonic tilting of rock strata provides the necessary condition for the formation of a confined aquifer.

3.5 Faults controlling groundwater flow

Faults cause discontinuities in rock sequences, and hence control groundwater flow. A fault may set high-conducting aquifer rocks against impervious rocks, resulting in water ascent along the fault zone and formation of springs, marking the fault line (Fig. 3.7). Occasionally, groundwater continues in its general flow path, across a fault, but in different rock beds (Fig. 3.8).

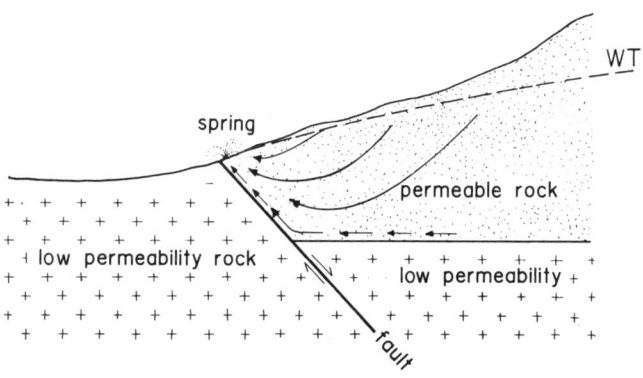

Fig. 3.7 A fault has placed a high-conducting aquifer against an igneous rock of low permeability. Water ascends along the fault zone, forming a line of springs.

Another fault-controlled example is the gathering of groundwater into rock beds of a rift valley (Fig. 3.9).

The above examples look simple, but in reality subterranean structures are often hard to recognize, and physical, chemical, and isotopic tracers are recruited to assist interpretation.

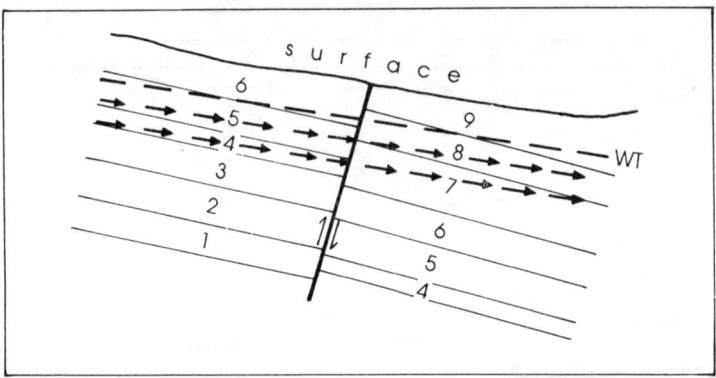

Fig. 3.8 Groundwater changing aquifer beds, due to strata displacement by a fault.

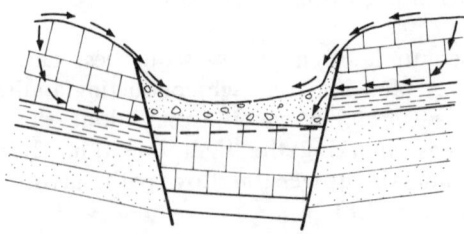

Fig. 3.9 Runoff and groundwater flowing into a rift valley.

Fault zones may be open, allowing rapid conduction of descending recharge water or ascending discharge water. In other cases fault zones may be sealed by mineral precipitation, forming subterranean barriers.

3.6 Intrusive bodies influencing groundwater flow

Igneous intrusive bodies are occasionally observed to act as subterranean barriers or dams. Good examples are dykes (Fig. 3.10), sills (Fig. 3.11), and stocks. Fresh igneous rocks are non-permeable, but with time they become fractured and may be somewhat conducting. Clay-rich weathering products may fill such fractures and improve the sealing properties.

In the case of dykes, water is dammed on the upstream side (Fig. 3.10). Hence, in placing a well near a dyke it is essential to:

- Recognize the existence of the dyke, which is often concealed by alluvium.
- Determine the direction of groundwater flow.

Lack of such information may result in a dry borehole, drilled on the downflow side of the dyke.

Geological Data 31

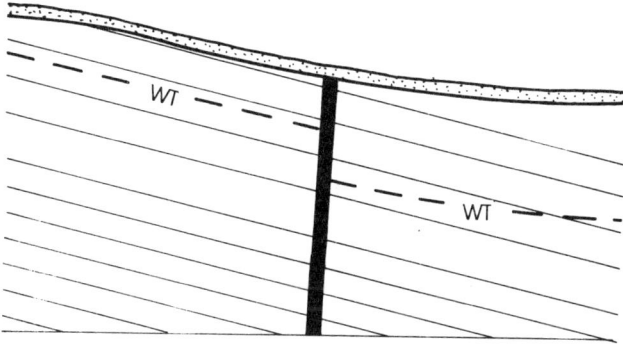

Fig. 3.10 A dyke intersecting the flow of groundwater, acting as a subsurface dam. The water table is higher on the upstream side, providing a promising site to drill a well.

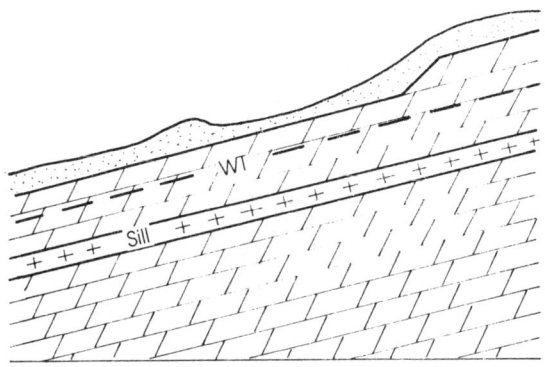

Fig. 3.11 A sill, intruded between carbonatic rock beds, forming a local aquiclude. Weathering into clay minerals may improve the sealing property of a sill.

4 PHYSICAL PARAMETERS

4.1 Water table measurements

The depth of the water table in a well is measured relative to an agreed mark at the top, e.g. the edge of the casing. This reference point should be clearly marked, to ensure reproducibility of the measurements. The depth of the water table is obtained in metres below the top of the well, or, more meaningfully, as metres below the surface, as shown in Fig. 4.1. AC is the depth of the water table below the reference mark at the edge of the well casing (or any other selected mark). This might slightly differ from BC, the depth of the water table below the surface. In the field notebook the depth AC should be used, as it is exact and properly defined. In a final report reference to the surface may be preferred. In any case, one has to specify how the depth is expressed: 'below well head' or 'below surface'.

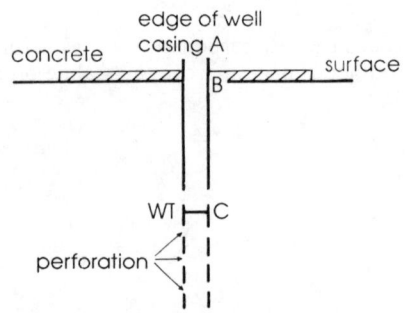

Fig. 4.1 Depth of water table: AC – depth below reference mark at the edge of the casing; BC – depth below surface.

A variety of simple, accurate, and professional tools is available on the market to measure water table depths to an accuracy of 0.2 cm. These instruments consist of a wire with a metal tip that is lowered, and upon hitting the water closes an electrical circuit, indicating the water table (Fig. 4.2.). The wire is marked for depth readings. Regional interpretation of water table data, obtained in several wells, is done in metres above sea level (masl). Hence, each well head has to be surveyed and its altitude has to be determined. The measured water table depth is subtracted from the well head altitude, in order to obtain the water table in absolute altitude units (Fig. 4.3). Water level measuring instruments are described in detail by Davis and DeWiest (1966) and Brassington (1988).

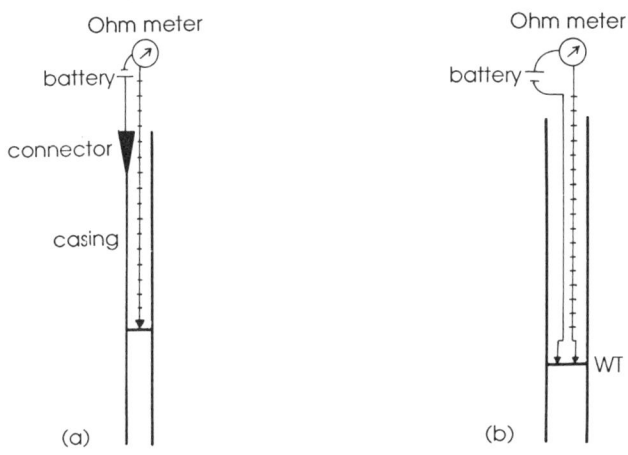

Fig. 4.2 Depth of water table (WT) measurement, based on the closing of an electrical circuit, by the water: (a) one wire, the casing serving as the second wire (provided it is made of metal and unbroken); (b) two wires lowered into the well.

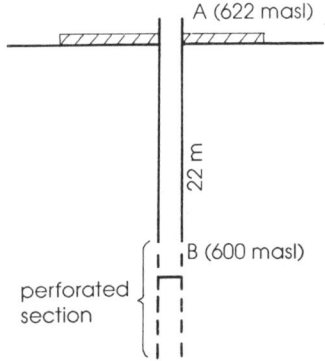

Fig. 4.3 Expressing water table in altitude units (relative to sea level): the measured depth AB is subtracted from the altitude of point A, determined by direct measurement, or read from a topographic map. In the example: the reference point A has been determined to be 622 masl (metres above sea level) and the depth of the water table was 22 m. Hence, the level of the water table in this well was 622–22 = 600 masl.

4.2 Interpretation of water table data

The depth of the water table is of interest to the management of the well and helps in identifying the aquifer rocks. The water table altitude is instrumental in the comparison of water tables of neighbouring wells, needed for the determination of the direction of water flow. Data may be processed as water table maps or as transects. These modes relate to a singular point in time, i.e. the regional water table as measured at a certain date. The next stage of interpretation deals with time variations of the water table, often referred to as *water table fluctuations*. The processing of water table measurements is dealt with in some detail, as it is an integral part of the processing of the hydrochemical and isotopic data.

Water table maps consist of contours of equal water table altitudes, as demonstrated in Fig. 4.4a. An immediate outcome is the deduction of the dominant direction of groundwater flow (Fig. 4.4b).

A certain degree of ambiguity, or subjectivity, is inherent in the compilation of water table data, as the solutions are not unique. However, in most cases the general picture is clear and the ambiguities are reduced as the number of measured wells increases.

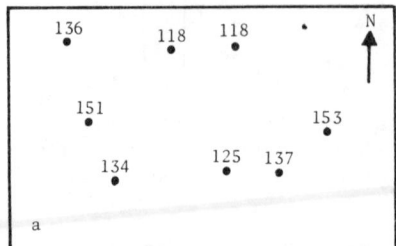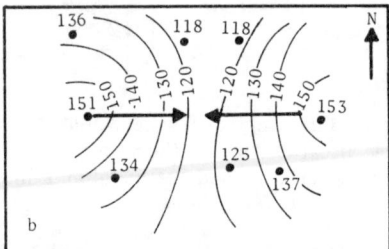

Fig. 4.4 Drawing equipotential lines: (a) map with well locations and water tables in masl; (b) equipotential lines based on the well data, and arrows showing *deduced* main directions of groundwater flow.

Water table transects portray the surface altitudes and water table altitudes of wells located along a line, as shown in Fig. 4.5c. The transect provides a convenient visual picture of the local topography, depth of groundwater, spacing of existing wells, and inferred direction of groundwater flow.

A water table transect may be superimposed on a geological transect, providing an insight into the groundwater system as shown in Fig. 4.6.

Water table fluctuations. The water table is dynamic in most systems. It changes in response to rain events, flood events, snow melt, recharge, and pumping. To decipher the interplay of these ongoing processes, periodic water table measurements are essential. Historical data are in many cases available from local water authorities, which conduct routine water table

Physical Parameters

Well no.	1	2	3	4	5	6	7	8	9
surface altitude (masl)	380	374	370	364	359	356	351	344	341
water table (masl)	346	344	341	339	337	334	332	330	330
distance on transect (km)	0	1.2	1.8	3.3	5.1	6.9	7.5	8.9	9.9

(a)

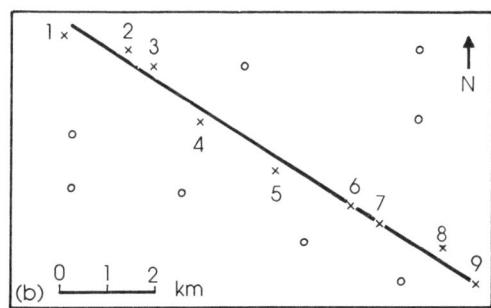

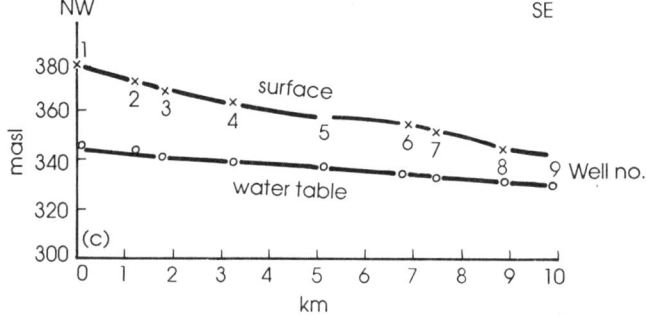

Fig. 4.5 Water table transect: (a) data table; (b) well location. The transect through the numbered wells (×) has roughly a NW–SE direction (o – other wells, not included in the transect). (c) the transect, drawn on millimetre graph paper.

measurements. Measurement of the depth of the water table at the time of each sample collection is essential, in order to couple the chemical and isotopic results with the hydrologycal data.

Repeated water table measurements in a well may be presented on a *hydrograph*, as a function of time. In the example given in Fig. 4.7, the low water table may be interpreted as reflecting lack of recharge in the winter and the rise may reflect snow melt recharge, followed by summer rains. Countless combinations of hydrograph shapes, and modes of their interpretation, are possible. Knowledge of local precipitation and climate is needed for proper interpretation. Rapid response to rain or flood events may indicate conduit-dominated intake (sections 2.1, 2.7, and 3.2), whereas response delayed by weeks or months indicates recharge through homogenous porous media.

Well no	depth (m)	lithology
1	86	soil (6 m) ; limestone (80 m)
2	90	soil (4 m) ; limestone (86 m)
3	75	soil (5 m) ; limestone (70 m)
4	75	soil (6 m) ; limestone (69 m)
5	46	soil (5 m) ; limestone (41 m)
6	72	soil (10 m) ; limestone (62 m)
7	54	soil (10 m) ; soil + gravel (6 m) ; calcareous sandstone (38 m)
8	58	soil (8 m) ; soil + gravel (14 m) ; calcareous sandstone (36 m)
9	59	soil (5 m) ; soil + gravel (18 m) ; calcareous sandstone (36 m)

(a)

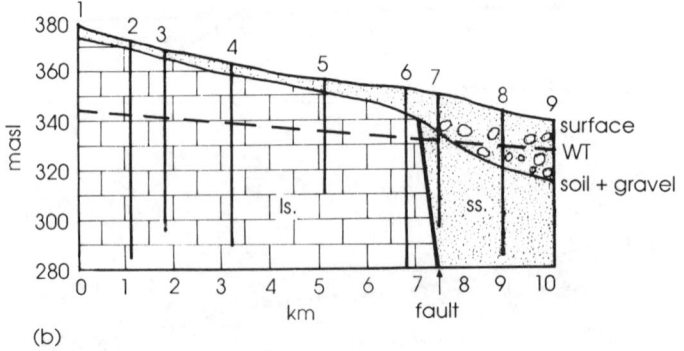

(b)

Fig. 4.6 Geological cross-sections: (a) data abstracted from geological reports, based on drill cuttings; (b) geological section (interpretation), along with the topographic (surface) and water table transects of Fig. 4.5.

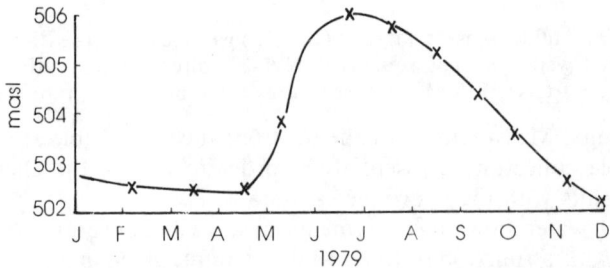

Fig. 4.7 A hydrograph based on repeated measurements of the water table in an observation well. Possible interpretation: winter months reflect restricted recharge, whereas from May onwards snow melt contributions are noticed, followed by summer rains.

Water table fluctuations are occasionally accompanied by measurable variations in water temperature or composition, providing crucial information on mixing of different water types (Chapter 6).

Recharge by different types of precipitation

A detailed hydrograph, based on frequent water table measurements at an observation well at the Saratoga National Historic Park, New York, USA, is given in Fig. 4.8, along with the local precipitation. The authors offered the following interpretation:

- Between January and May snow melt provided some recharge, resulting in a slow rise of the water table.
- In April the rainy season began, augmenting the continuing snow melt and causing a further rise of the water table.
- In June the principal growing season began, causing increasing transpiration losses that lowered the water table.
- 4.7 inches of rain fell during a four-day hurricane in September and the water table rose slightly.
- During the cold month of December only snow fell and the water table continued to decline.

This detailed tracing of recharge history, reflected in the water table fluctuations, is unique in its completeness and reflects the direct local

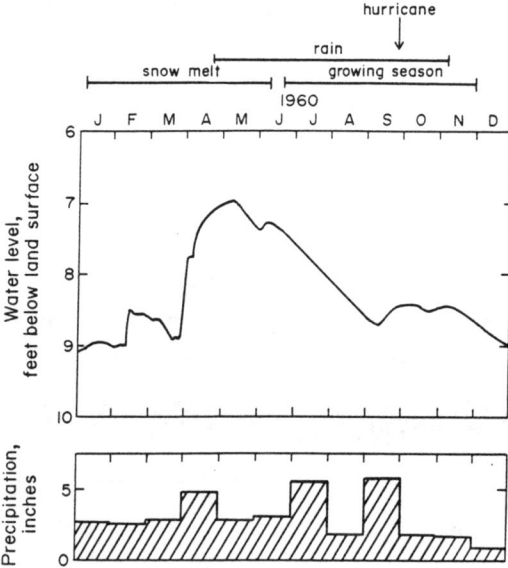

Fig. 4.8 Hydrograph of an observation well at the Saratoga National Historic Park, New York, USA (Winslow *et al.*, 1965) and local precipitation. Interpretation is discussed in the text.

recharge and the lack of pumping in the Saratoga Historic Park terrain. This case study demonstrates how rewarding a simple set of data can be, a topic to be further discussed in section 7.6 in relation to time-data series.

Recharge by a river

Recharge of a well by the Mohawk River, New York, USA, has been demonstrated by Winslow *et al.* (1965) by studying hydrographs of the well and the river (Fig. 4.9):

- The water table of the well followed the river during August.
- In September a sudden rise in the river was caused by a flood event, followed by an immediate rise of the water table in the well.

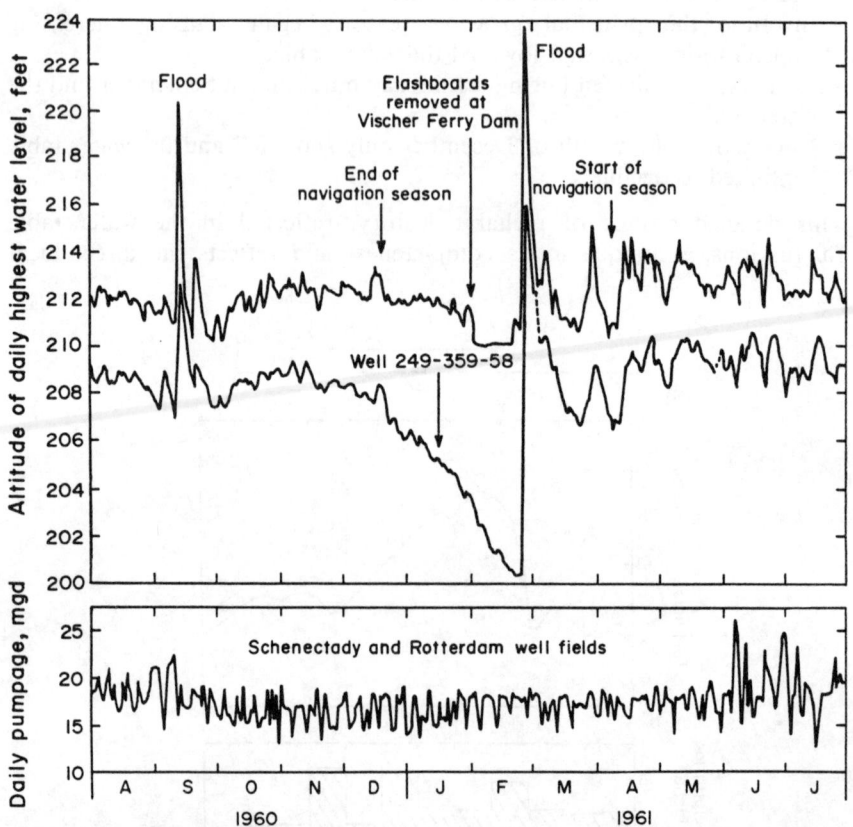

Fig. 4.9 Hydrographs of the Mohawk River and an adjacent observation well (Winslow *et al.*, 1965). The perfect match (see text) proves recharge from the river. Pumping from adjacent wells (bottom) was steady and could not cause the observed changes of the water table in the well.

- Lowering the river level by the opening of a dam, at the end of the navigation season (December), was followed by a sharp decline in the well.
- A flood, at the end of February, caused a sharp rise in the well.
- Stabilization of the river level, at the beginning of the new navigation season, was followed by a parallel stabilization in the well.

The close response of the well to fluctuations in the river level demonstrates that the well is dominantly recharged by the Mohawk River. Possible interferences by pumping of adjacent wells may be ruled out on the basis of the monotonic abstraction, portrayed at the bottom of Fig. 4.9.

Velocity of a recharge pulse

Changes in the water table of the Mohawk River and a number of adjacent observation wells are reported in Fig. 4.10, taken from Winslow et al. (1965). The wells followed the river, with a time lag of 4–12 hours (insert in Fig. 4.10). Two possible explanations for this time lag may be envisaged: arrival of the hydraulic pulse, or arrival of the recharge front (assuming piston flow) (section 2.9). To tell the two apart, the time lag observed for these wells by temperature measurements is helpful (Fig. 4.21), as discussed in section 4.8. The temperature time lag of, for example, well 58, has been observed to be about 3 months, whereas the water table time lag was only 12 hours. Hence, the latter defines the arrival of the hydraulic pulse, whereas the former defines the travel time of the recharge front. The distances given

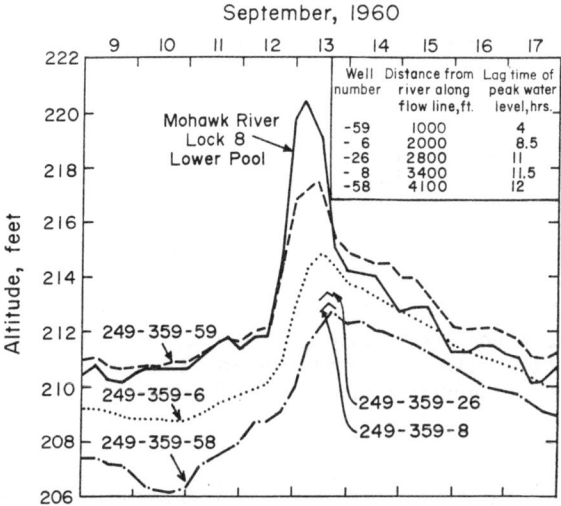

Fig. 4.10 Water table fluctuations measured at wells near the Mohawk River (Winslow et al., 1965). The wells followed a flood event with a time lag of several hours, correlated with the distance from the river (insert). The distance, divided by the time lag, provided the propagation velocity of the hydraulic pulse.

in the insert in Fig. 4.10, divided by the respective time lags, provided the propagation velocity of the hydraulic pulse, which was 300 ± 50 ft/h, or 100 ± 20 m/h.

The case studies discussed above, and depicted in Figs 4.8–4.10, reveal the importance of repeated measurements, providing evolution with time, and also the importance of auxiliary data, such as distribution of local precipitation, discharge in adjacent pumping wells or repeated temperature measurements.

4.3 Gradient and flow direction

The gradient, or drop in elevation, of the water table along the flow path of the water is a major factor in defining the water flow velocity, as derived from Darcy's law (section 2.8).

The gradient is expressed in metres drop of height per kilometre of horizontal flow trajectory, as shown in Fig. 4.11. The water table gradient is calculated from specific water table measurements or extracted from a water table map, as shown in Fig. 4.12.

The direction of flow of groundwater is deduced from the observed water table gradients. Water flows mainly along the maximum gradient, as shown in Fig. 4.4b. It is essential to remember that by this procedure one finds only a suggested flow direction; whether it is real has to be checked by other observations, as discussed in the following section.

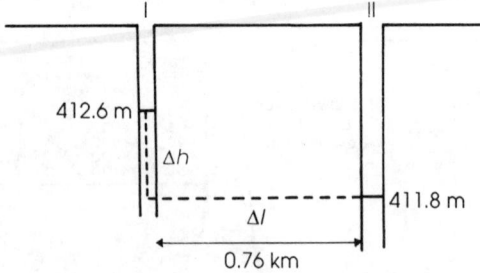

Fig. 4.11 Determining flow gradient. The drop of water table between wells I and II is $\Delta h = 412.6 - 411.8 = 0.8$ m; the distance (read from a map) is 0.76 km. Hence, the hydraulic gradient is $0.8/0.76 = 1.0$ m/km.

4.4 The need for complementary data to check deduced gradients and flow directions

Determination of the flow gradient between two wells (Fig. 4.11) is based on the assumption that the two wells are hydraulically interconnected, as

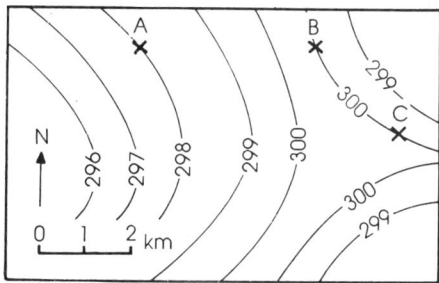

Fig. 4.12 Extracting flow gradients from a water table map (contours in masl). Δh, from B to A, is $300.0 - 298.0 = 2.0$m, and the distance (read from the map) is 3.7 km. Hence, the gradient between B and A is 0.54 m/km. The gradient from B to C is 0 m/km.

shown in Fig. 4.13a. However, wells may be separated by an impermeable rock bed, with no hydraulic interconnection (Fig. 4.13b). Thus, the gradient measurement between the two wells in the second case is meaningless. The same holds true for deduced direction of water flow – it is meaningful, provided the involved wells are hydraulically interconnected. Water flows from I to II in Fig. 4.13a, but it does not flow from III to IV in Fig. 4.13b. In addition, a variety of processes lowers the local water conductance, occasionally preventing lateral flow. An example for such a process is chemical clogging (Goldenberg et al., 1983).

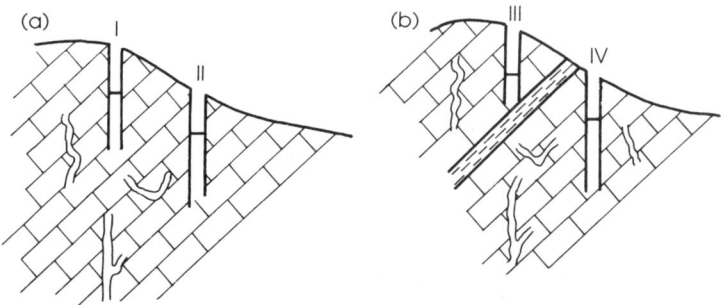

Fig. 4.13 Hydraulic interconnection between wells in fissured limestone terrain: (a) wells I and II are interconnected and water flows down gradient. (b) similar looking wells, separated by an impermeable rock bed. In this case wells III and IV are not interconnected, in spite of the *apparent gradient*.

Thus, other parameters are needed to check for hydraulic interconnections suggested between studied wells. Chemical constituents of the water may indicate whether studied wells tap the same water type or belong to different water groups. In the latter case simple hydraulic interconnections may be ruled out (section 6.5). Temperature of groundwater commonly increases down flow and, hence, if colder water is encountered along a suggested flow path, straightforward interconnection between wells is disproved (section 4.8). The isotopic composition of water provides additional clues to hydraulic interconnection (section 9.10) and so do the radioactive age indicators tritium (section 10.7) and carbon-14 (section 11.9).

4.5 Velocities and pumping tests

As stated in section 2.8, the velocity by which groundwater flows is commonly calculated from the water table gradient and the coefficient of permeability, K (or the related parameter of transmissivity). K is determined by a pumping test. During such a test a studied well is intensively pumped, and the water table in one or two adjacent observation wells is monitored, as well as the drop in the pumped well. The change of the water table level as a function of the pumping rate serves to compute the aquifer permeability.

Pumping tests call for expertise that is beyond the scope of the present section, but as shown in the next section, the incorporation of hydrochemical methods is essential for safe interpretation of pumping test data.

For a detailed discussion of the theory of groundwater flow the reader is referred to the book of Davis and DeWiest (1966).

4.6 Chemical and physical measurements during pumping tests

The interpretation of pumping test data (Todd, 1980; Walton, 1988) is based on the assumption that only one aquifer is pumped and tested. However, the intensive pumping during the test causes a significant local pressure drop in the pumped aquifer that may cause water of an adjacent aquifer to breach in (Fig. 4.14). If the pumping test is done in a phreatic aquifer, water of a lower, confined, aquifer may flow in. Similarly, in pumping tests in confined aquifers, an overlying phreatic aquifer may be drawn in. Leaky aquifers may be detected, or suspected, from the test results, but no hint is gained as to the nature of the invading water, neither the amounts involved. The breaching in of water from a second aquifer has to be detected, so that it can be included in the pumping test interpretation. Otherwise, the conclusions based on the test may be erroneous: apparent conductivities that are too large may be deduced, and exaggerated operation pumping rates may be suggested.

To obtain the necessary information on the number of aquifers effectively included in a pumping test, continuous measurements of temperature, conductivity, and other parameters are recommended before, during, and after the test. Temperature is sensitive to aquifer depth (section

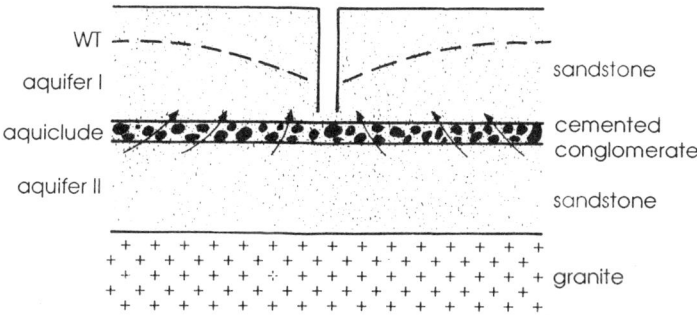

Fig. 4.14 Possible effect of a pumping test: the water table of aquifer I is drawn down near the well, a feature called a *depression cone*. Occasionally, water from a lower confined aquifer may breach in (arrows across the aquiclude).

4.7) and most useful in distinguishing waters originating from different depths. A constant temperature value, as shown in Fig. 4.15a, is a favourable indication that the pumping test remained restricted to a single aquifer. In contrast, in the example given in Fig. 4.15b, warmer water intruded into the pumped aquifer and the later part of the pumping test included water from two aquifers.

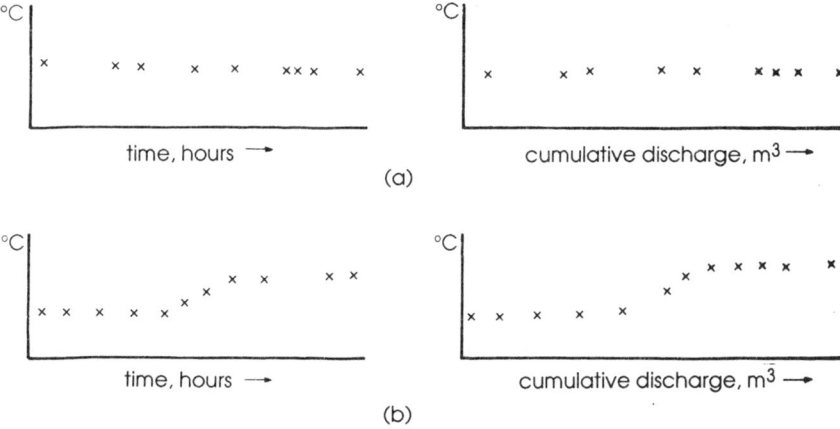

Fig. 4.15 Temperature measurements during pumping tests, expressed as a function of time and cummulative discharge. (a) temperature remained constant, indicating pumping remained restricted to a single aquifer. (b) temperature suddenly increased, indicating water from a warmer (probably lower) aquifer breached in.

Electrical conductivity is readily measured in the field and reflects the total amount of salts dissolved in the water (section 8.4). It is highly recommended that electrical conductivity is measured at close intervals through a pumping test. The data may be plotted against time or, preferably, against cumulative abstraction. In the example of Fig. 4.16 it is seen that the intruding water is saltier.

This result may be of importance for the regular pumping rate to be decided for the well. If the additional salinity is not harmful, then a relatively high abstraction rate may be recommended; but if the additional salinity is unacceptable, then a low rate of routine abstraction has to be

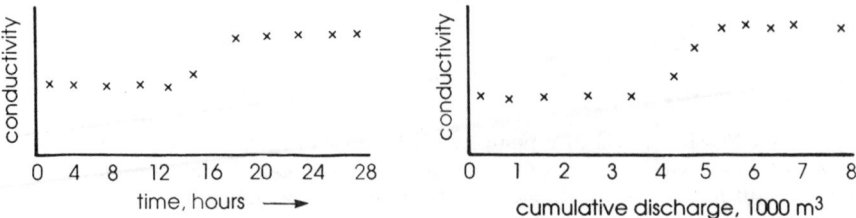

Fig. 4.16 Conductivity measurements during a pumping test - saltier water intruded the pumped system after 12 hours, during which 3500 m³ were pumped.

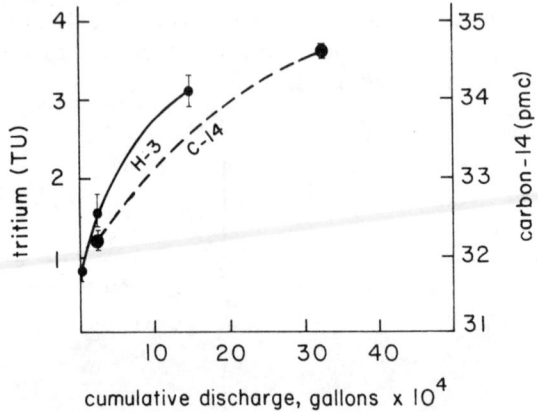

Fig. 4.17 Tritium and ^{14}C measurements during a pumping test, conducted in a confined aquifer at the Aravaipa Valley, Arizona, USA (Adar, 1984). Recent water, of high tritium and ^{14}C concentrations, intruded from the overlying phreatic aquifer.

adapted. In such considerations the presentation of measured parameters as a function of both time and accumulative discharge is most useful. In the example of Fig. 4.16 the salty water broke through after pumpage of 3500 m³ in 12 hours. Hence, the routine pumping rate should be significantly lower if only fresh water is wanted, and abstraction may be closer to the break-through value if water quantity is of prime importance and the quality of the mixed water is acceptable.

In addition to the direct field measurements, water samples should be collected for possible detailed chemical and isotopic analysis. As a rule of

Physical Parameters 45

thumb, a pumping test may be accompanied by one sample taken before the test, representing the non-disturbed water system, and 10 samples periodically collected during the test. It is then suggested that the first and last samples are fully analysed. If they turn out to be alike, and no temperature and electrical conductivity changes have been noticed, then one may confidently conclude that the pumping test affected only one aquifer. If the first and last samples turn out to differ from each other, then the rest of the collected samples should be analysed to define the nature and mixing percentages of the waters involved. The laboratory analyses are costly, but the information derived is of prime importance. Failure to notice the intrusion of a second water type may turn out to be a failure to predict possible deterioration of water quality due to overpumping of a well system. A case study of a documented pumping test is revealed in Fig. 4.17, taken from a study in the Aravaipa Valley, Arizona. The outcome of a fully documented pumping test is a corner-stone of the hydrochemist's report (Chapter 14).

4.7 Temperature measurements

Temperature field measurements. Commercial thermometers (mercury or digital) are adequate for most hydrological studies, their readings being good to $\pm 0.5°C$. However, they should be calibrated in the laboratory, as they are occasionally off by up to $\pm 1°C$. Calibration may be carried out by reading the thermometer in a fresh water–ice mixture, providing $0°C$, and in boiling water, providing the local boiling temperature (determined by the local altitude). The required corrections should be written on the instrument and applied to the field readings. It is recommended always to use two thermometers, as a check and as a spare in case of breakage.

Temperatures should be read while the thermometer is dipped in the water, because upon taking it out the reading drops rapidly during drying. If reading while the thermometer is dipped in the water is not feasible, a bottle may be filled with the water and the thermometer dipped in it. The cooling of a bottle of water is not too fast.

The temperature of the water of a spring should be taken as deep as possible. If the water pool of a spring is large, several readings should be taken to reach the extreme value, i.e. warmest or coldest, most different from ambient air temperature.

If the temperature measured at a spring is close to the ambient air temperature, temperature re-equilibration of the near-surface spring water could have occurred. A second measurement, at night or at a different season, may clear up this point. If the same value is repeated, in spite of a variation in the ambient air temperature, then the measured spring temperature is the true one, reflecting the temperature at depth.

Depth of groundwater circulation, deduced from temperature data. Temperature measured in wells and mines reveals a general increase with depth, or a geothermal gradient (Fig. 4.18). The value of the geothermal gradient varies from one location to another. An average value often quoted is 3°C/100 m.

Groundwater is commonly temperature-equilibrated with the aquifer rocks. Thus, temperatures measured in springs or wells reflect the temperature attained at depth, and therefore provide information on the depth of circulation. The calculation is straightforward:

$$\text{depth (m)} = \frac{T_{\text{measured}} - T_{\text{surface}}}{\Delta T/100}$$

where: T_{measured} is the measured spring or well temperature, T_{surface} is the local average annual surface temperature, obtained from climatological maps or local meteorological institutes, and ΔT is the local geothermal heat gradient, established by geophysical studies.

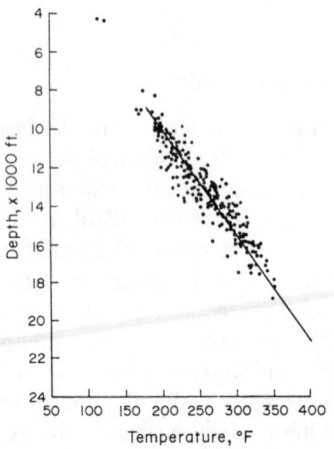

Fig. 4.18 Temperature measurements in 48 deep wells in south Louisiana (from Bebout and Guttierrez, 1981). A gradient of 1.8°F/100 ft, or 3°C/100 m, is observed.

Example. A spring of 62°C emerges in a region with an average annual surface temperature of 22°C. The local heat gradient has not been measured, hence the value of 3°C/100 m may be applied. The depth of circulation is:

$$\frac{62 - 22}{3/100} = 1330 \text{ m}$$

Such values are minimum depth values, as the water may cool during ascent. The larger the water flux, the less cooling there is. As a rule of thumb (which warrants quantitative checking), for a spring or well with a discharge of 30m³/h, or more, cooling is negligible.

For cases of lower discharge rates the calculated circulation depths are to be regarded as minimum values.

The reconstruction of the depth of groundwater circulation may seem trivial, but temperature is the best and most reliable tool to establish it. Knowledge of the depth of circulation may be further used to infer the type of rocks in the geological column passed, and the area of recharge, topics to be discussed later on in the light of case studies.

Type of groundwater flow traced by water temperature measurements. Recharge water often has a temperature that differs significantly from the aquifer temperature. Continuous temperature measurements may, thus, serve as excellent recharge indicators. Figure 4.19 shows biweekly temperature measurements, conducted by Shuster and White (1971) in springs of two regions of carbonate rocks in the central Appalachians. The researchers recognized, by independent observations, two types of springs:

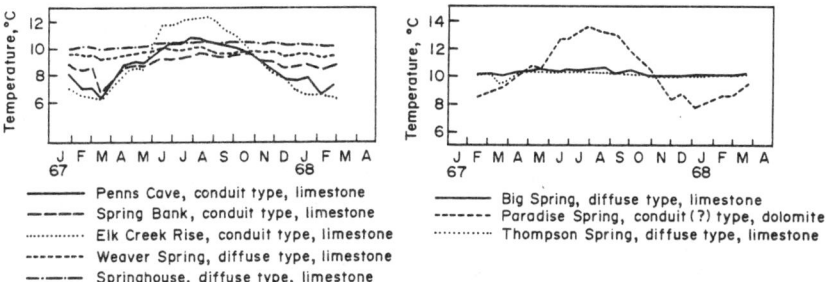

Fig. 4.19 Seasonal temperature observations in springs of the Penns Greek Basin, central Appalachians, USA (from Shuster and White, 1971). The authors classified the springs (see text) into a conduit type with rapid karstic flow, and a diffuse type with slow flow through minute rock fractures. The former revealed significant seasonal temperature variations, whereas the latter manifested a steady temperature.

- A conduit type of rapid karstic flow, recognized by adjacent sink holes and by their muddy, or turbid, water.
- A diffuse type, with slow flow through minute fractures in the rock, recognized by the clear water and lack of adjacent karstic morphology.

The temperature records in Fig. 4.19 correlate excellently with this classification. The conduit type revealed significant temperature variations, in good agreement with the known seasonal variations of karstic flow, whereas springs with the diffuse type of flow revealed steady temperatures. Once the temperature pattern has been established, by calibration against other observations, the tool could be applied to the Paradise Spring (Fig. 4.19), defining it as a conduit type, although no supporting field data were available.

Mixing of groundwaters may occasionally be detected by temperature measurements, e.g. drop of temperature caused by the arrival of snow melt recharge, seasonally added to a regional base flow. Temperature measurements are most useful in detecting intermixing when coupled with chemical and isotopic measurements, a topic discussed in sections 6.6 and 6.7.

4.8 Tracing groundwater by temperature: a few case studies on the Mohawk River

A large-scale temperature study was reported by Winslow *et al.* (1965), studying recharge from the Mohawk River into gravel deposits underlying the river flood plain. They monitored temperatures in well fields with a total yield of $7 \times 10^4 \text{m}^3/\text{d}$. The temperature data tracing the river recharge have been worked out in a number of ways, all providing superb examples for processing temperature data.

Contours of equal groundwater temperature. Temperature data are occasionally presented in contours of equal groundwater temperature (results of measurements in a large number of wells). These contour maps (Fig. 4.20) are given for twelve successive months during 1960/61. The river temperature (written in the left lower corner of each monthly map) is seen to decrease from 64°F in the autumn down to 32°F (freezing) in the winter. In the spring it warms up, to reach 77°F in the summer. Thus, twice a year the temperature gradient between the river and the aquifer is reversed. The temperatures in the aquifer are seen to respond to these changes, but with a certain time lag. In October the temperature in the aquifer was in the range of 70°F near the river and 50°F 900 m south, as compared to 64°F in the river. This range dropped to a minimum of 40° near the river and 50°F 900 m away, in March. By that time the river had already started to warm up. The aquifer reached maximum temperatures of 75°F near the river and 50°F 900 m away in September, when the river was 77°F. Another look at the October map completes the cycle. These significant variations in the aquifer temperature, lagging behind the changes in the river, prove that the aquifer is recharged by the river. The changes are greater near the river (35°F to 75°F) and are damped away from it until at a distance of 900 m to the south the temperature is steady (50°F) throughout the year. This lateral temperature gradient points again to recharge from the river.

Seasonally repeated depth profiles in a well. A second way of presenting the data of the Mohawk River temperature survey is in the form of seasonally repeated depth profiles in a well (Fig. 4.21). The temperatures are seen to increase for half the year and then decrease, proving recharge, similar to the mode seen in the temperature maps (Fig. 4.20). The profiles show the vertical dimension of the recharge: temperature fluctuations are accentuated between 180 and 200 feet. Recharge is most efficient in this horizon, indicating that it has the highest conductance.

Physical Parameters

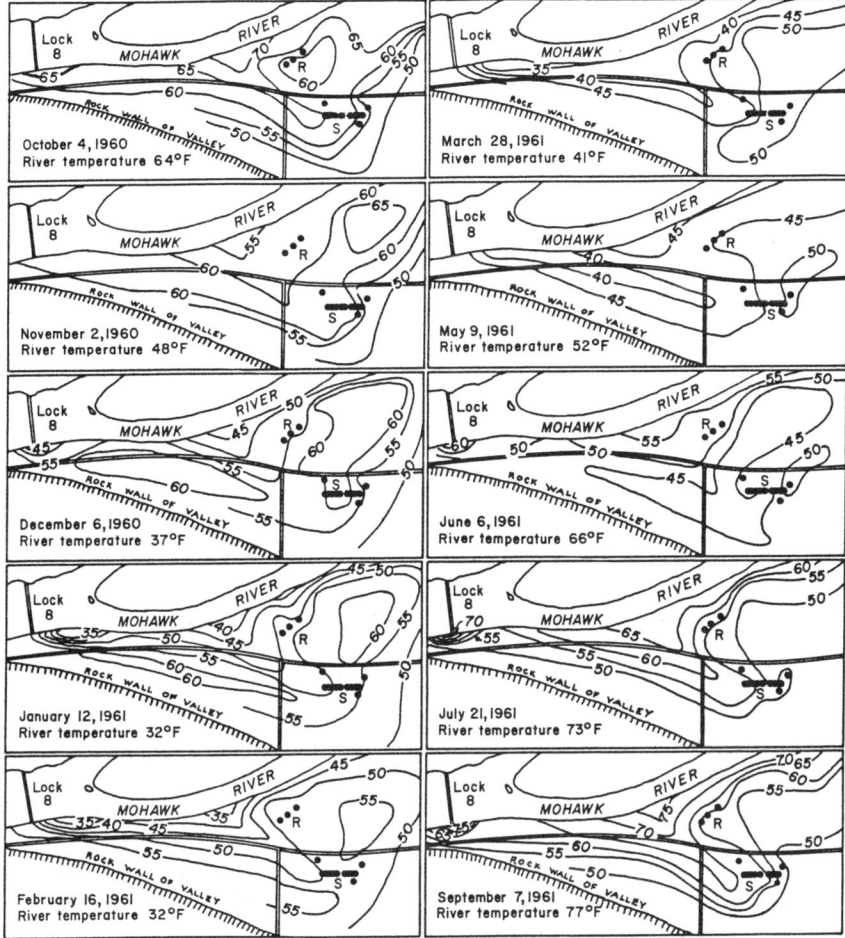

Fig. 4.20 Monthly maps of equal groundwater temperature for a well field boardering the Mohawk River (from Winslow *et al.*, 1965). The river temperature changed over an annual cycle from 77°F to 32°F (lower left corner of each map). The aquifer followed these temperature changes (temperature line closest to the river), indicating recharge from the river.

Map of contours of the annual range of temperature variations. Winslow *et al.* (1965) also expressed the results of their temperature study as a map of contours of the annual range of temperature variations.

The recharge from the river and the movement southward are well seen in the southward damping of the range of annual temperature variations. A detail clearly verified is that of a zone of preferred conductivity, marked A in Fig. 4.22.

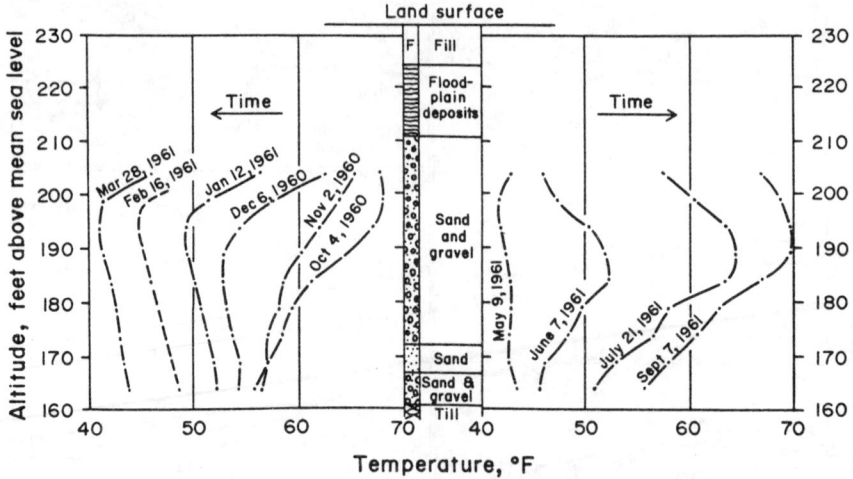

Fig. 4.21 Seasonal temperature profiles in well 61, 100 m away from the Mohawk River (from Winslow *et al.*, 1965). The temperatures are seen to decrease from October to March and to increase from June to September, like the trends seen in the temperature maps of the previous figure. The profiles reveal that the largest temperature variations occurred at a depth interval of 180-200 ft above sea level, indicating recharge occurred mainly through this part of the rock section, which in turn, must therefore have a higher lateral conductivity.

Temperature time-data series. A fourth way in which temperature data from the Mohawk River were processed is shown in Fig. 4.23, in which temperature observations over a whole year are plotted for six wells and for the river. Wells 54 and 59 are seen to follow the pattern of the river, but with a time shift of about 2 months. Well 21, in contrast, reveals no resemblance to the river at all and the rest of the wells have intermediate patterns. The degree of similarity to the river is correlated with the distance of the well from the Mohawk, or, as more precisely stated by the researchers, the degree of similarity reflects the hydraulic distance. By this term they mean the combined effect of distance, conductivity, amount of water in transit, and temperature equilibration with the aquifer materials.

The temperature graphs in Fig. 4.23 serve not only to demonstrate recharge. Effective water flow velocities are obtained from the distances between each well and the river, divided by the time lags in the response of each well to the temperature changes in the river. A treatment of this kind for all wells showed up a zone of high velocities, marked A in Fig. 4.22.

4.9 Cold and hot groundwater systems

Groundwater temperatures vary from a few degress above freezing to boiling, and in geothermal wells fluids with temperatures over 300°C are

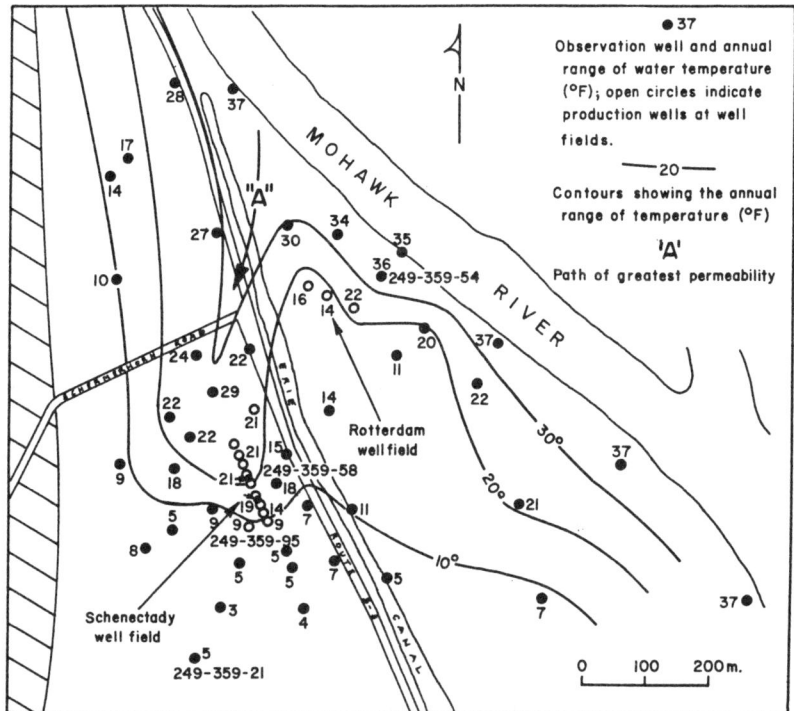

Fig. 4.22 Location map of wells near the Mohawk River with values of the range of annual changes in groundwater temperature (from Winslow *et al.*, 1965). Using these values, contours of equal annual temperature variation were drawn. The decrease of these contour values indicates water moves from the river into the aquifer, especially through the zone marked A.

exploited. This huge range of temperatures is formed by a variety of discharge mechanisms and, to a large extent, by differences in the depth of circulation and local heat gradient values.

Temperature data closely reflect hydrological conditions and several types of groundwater may accordingly be distinguished:

Groundwater colder than local annual surface temperature occurs in high altitudes and is caused mainly by snow melt recharge.

Groundwater with temperature close to local average annual surface temperature belongs to the shallow active water cycle, circulation being limited to 100 or (rarely) 200 m.

Groundwater with more than 6°C above local average annual surface temperature circulates to appreciable depths, deducible from the heat gradient (section 4.7).

Warm springs with temperatures up to 65°C are common in tectonically bisected terrains, where groundwater can circulate to appreciable depths.

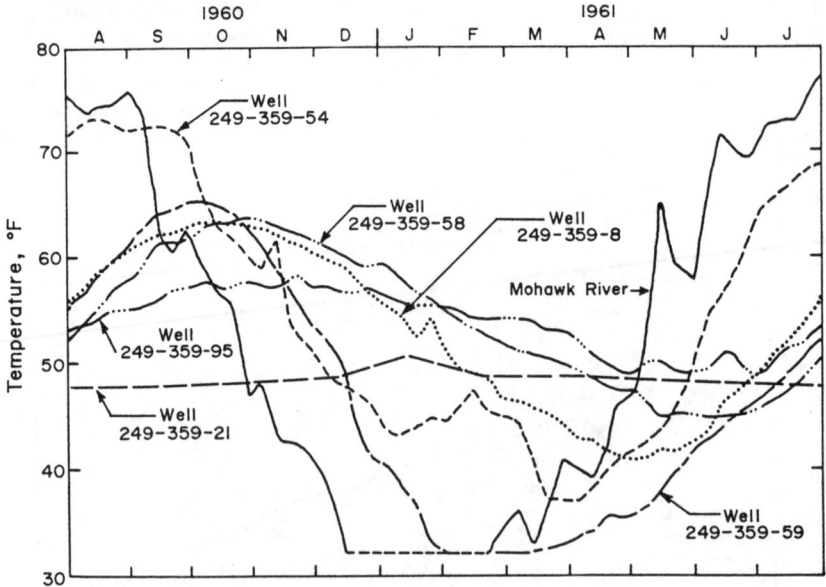

Fig. 4.23 Temperature time-data graph of wells near the Mohawk River (Winslow *et al.*, 1965). Wells 54 and 58 follow the river temperature changes with a time lag, applicable to calculate the velocity of the recharged water. Well 21, most distant from the river, revealed a steady temperature over the year, indicating the river recharge is probably not contributing to this well. The rest of the wells showed intermediate degrees of temperature response to the river temperature variations, in proportion to their hydraulic distances.

Most graben, or rift, systems and active orogenic regions host hot springs explained in this way. Certain large synclines, or sedimentary basins, contain huge amounts of warm groundwater, exploitable for space heating. The Paris basin is an example.

Geothermal systems host boiling springs, and in them boreholes encounter steam. Classical examples are known from the USA (e.g. Yellowstone), Italy (e.g. Pisa), Iceland, and New Zealand. In recent decades methods of harvesting the energy of geothermal systems have advanced greatly, and a whole field of geothermal prospection and production has evolved. Geothermal manifestations are closely correlated to recent magmatic activity and movement of crustal plates. The associated waters are concentrated brines, rich in CO_2.

The topic of geothermal systems will not be dealt in the present book, but as hydrological systems their methods of study overlap those of nonthermal groundwater.

4.10 Discharge measurements and their interpretative value

Discharge is the measure of the amount (volume) of water emerging in a spring, or pumped from a well, per unit time. A large number of units is used for discharge in the literature, but m³/h (cubic metres per hour) is most recommended. Discharge may, in certain cases, be measured in the field with the aid of a container of known volume and a stopper, or a watch with marked seconds. The discharge is calculated from the volume (number of times a vessel has been filled) and the time involved. Occasionally spring water flows from several directions, the output is too large, or the well discharges into a closed system of pipes. In such cases the discharge information has to be obtained from the well operator or local water authority.

Discharge of a spring or a well is a most informative parameter, as it provides insight into the quantitative aspect of groundwater hydrology.

Figure 4.24 includes monthly measurements of discharge, temperature, and chlorinity of a spring. Constant discharge is observed, indicating that water is delayed in the aerated zone, flowing through a porous medium (in conduit-dominated recharge seasonal fluctuations are reflected in

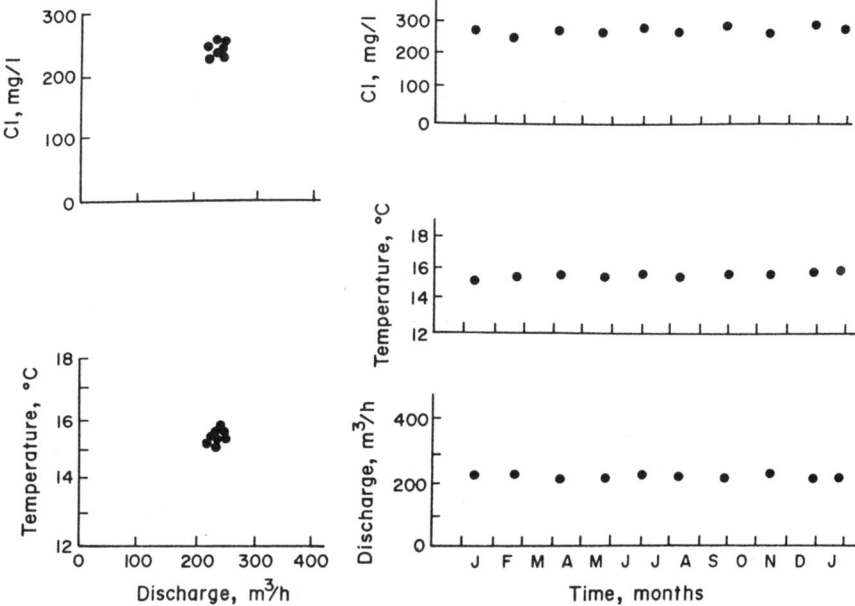

Fig. 4.24 Monthly measurements of discharge, temperature and Cl concentration in a (hypothetical) spring. The three parameters are constant over the year, indicating only one type of water is involved, recharged through a porous medium (non-karstic), and the system's storage capacity is large compared to the annual recharge and discharge.

variations in spring discharge). In addition, constant discharge indicates that the water storage capacity of the system is large, compared to the annual recharge or discharge. The accompanying temperature and Cl values are also steady, indicating that one type of water is involved in the system. Figure 4.24 manifests two useful models of presentation of recharge data: as a time-date series and as a function of other measured parameters. The conclusion that one type of water is involved can be deduced from the horizontal line in the time-data diagram or from the narrow cluster of values in the parametric diagrams.

Data plotted in Fig. 4.25 are from a different case, but resemble the parameters reported in Fig. 4.24. The pattern obtained is different: the discharge varied in an annual cycle, but the temperature and chlorinity remained constant in this case also. Thus, the interpretation is again of one type of water being involved, but recharge is via conduits of a karstic nature. Temperature is close, in this case, to the average annual surface temperature and fluctuations in the recharge water temperature are damped by intermixing in the aquifer. Thus, the storage capacity of the system is large as compared to the peak recharge.

A third example, given in Fig. 4.26, reveals a case in which recharge varied seasonally in a spring and temperature and chlorinity varied as well, but in an opposite pattern (right diagrams in Fig. 4.26). The data plot along

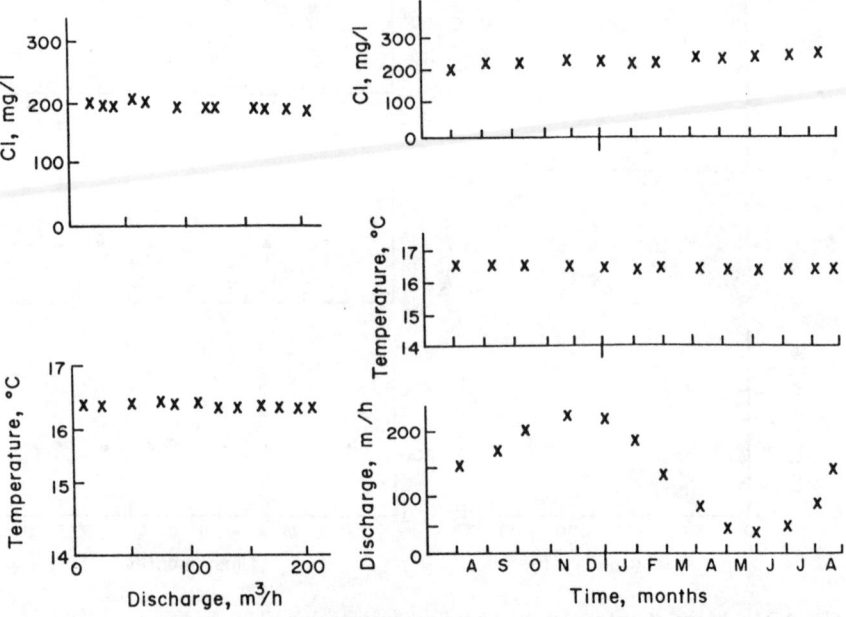

Fig. 4.25 Monthly measurements of discharge, temperature and Cl in a (hypothetical) spring. Temperature and chloride are constant, revealing that one type of water is involved. The significant discharge variations indicate recharge is fast and of a karstic nature.

straight lines in the left diagrams of Fig. 4.26, revealing a negative correlation between recharge and temperature or chlorinity. Such correlation lines indicate intermixing of two water types, a topic fully addressed in section 6.6. A warmer and more saline type of water intermixes (in varying percentages) with a colder and less saline water. The highest possible temperature of the warmer end member might be deduced by

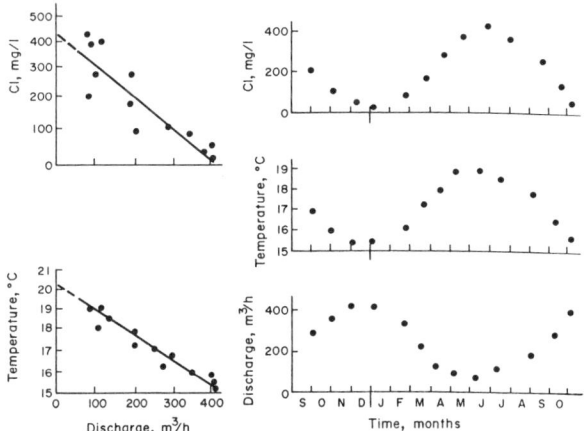

Fig. 4.26 Monthly measurements in a (hypothetical) spring. Discharge, temperature, and Cl varied considerably, indicating that (a) more than one type of water is involved (e.g. base flow and seasonal additions); (b) recharge is karstic. The data plot along straight lines in the temperature–discharge and Cl discharge graphs, indicating that two end members are intermixing in the spring system. The temperature and Cl values of the warm and more saline endmember may be deduced by extrapolation of the best-fit lines to zero discharge (see text).

extrapolation of the best-fit line to zero discharge on the temperature–discharge graph. A value of 20.3°C is obtained. In a similar way, the highest possible value of Cl concentration of the warm end member may be deduced by extrapolation to zero discharge in the Cl discharge diagram in Fig. 4.26. A value of 420 mg Cl/l is obtained. As zero discharge has no meaning in our context (a dry spring has no temperature or Cl concentration), it is clear that the true intermixing water end members have values that lie between the extrapolated values (20.3°C, 420 mg Cl/l) and the highest observed values (19°C, 400 mg Cl/l). In the present example the two sets of values are very close. The topic of water mixtures is discussed in sections 6.6 and 6.7. Discharge measurements often provide negative correlation lines, essential in finding end member properties.

A case study from the Yverdon spring, western Switzerland (altitude 438 masl) is summarized in Fig. 4.27. Maxima in the discharge curve are seen in general to be accompanied by minima in the temperatre and electrical conductivity (reflecting salinity) curves, indicating that a warm and more saline water intermixes with a colder and fresher water. The discharge and temperature data of the Yverdon spring have been replotted in Fig. 4.28. The best-fit line, extrapolated to zero discharge, intersects the temperature axis at 30°C. This is a maximum possible temperature of the warm end member in the Yverdon complex. The true end member temperature is between this extrapolated value (30°C) and the highest observed temperature (24.3°C).

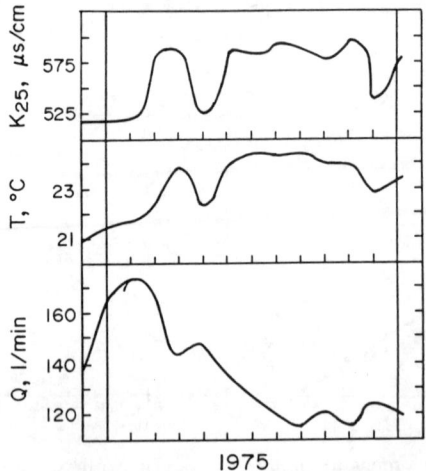

Fig. 4.27 Discharge, temperature, and conductivity (reflecting salinity) in a spring at Yverdon, western Switzerland (Vuataz, 1982). Temperature and conductivity are mirror images of the discharge, revealing a negative correlation to discharge (see also next figure), indicating mixture of two end members (see text).

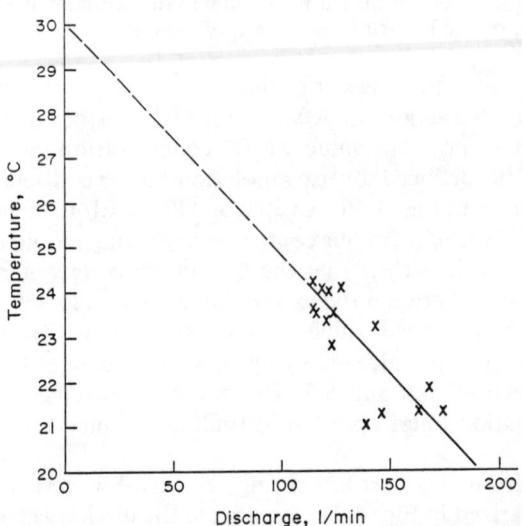

Fig. 4.28 Temperature-discharge data (from previous figure) of the Yverdon Spring, Switzerland. The negative correlation enables one to deduce an upper limit for the temperature of the warm end member (30°C), by extrapolating to zero discharge (see text).

5 ELEMENTS, ISOTOPES, IONS, UNITS, ERRORS

The outcome of hydrochemical studies is based on chemical data and their interpretation. A clear knowledge of basic terms, units applied, and the nature of analytical errors, is essential to avoid confusion and mistakes. The following concepts and definitions have to be mastered as a prerequisite to data processing.

5.1 Elements

To define an element one has to discuss the basic units of nature – the atoms. Atoms are built of a nucleus that consists of two main types of particles (nuclides) – protons, which have a positive charge, and neutrons, which are neutral. Around the nucleus orbit much smaller particles with a negative charge, called electrons. In each atom the number of orbiting electrons matches the number of protons in the nucleus, balancing the negative and positive charges and making the atom neutral. The chemical properties of atoms are defined by the number of electrons, which in turn, are defined by the number of protons in the nucleus.

Chemical elements are each made of only one type of atoms, i.e. atoms with the same number of protons. For example, hydrogen atoms contain one proton, helium atoms contain two protons, oxygen atoms contain eight protons and uranium atoms contain 92 protons (Table 5.1).

5.2 Isotopes

The number of neutrons in an atomic nucleus either equals the number of protons or is slightly greater. Most elements have several types of atoms, differing in the number of neutrons that accompany the protons, whose number is fixed for each element. For example, hydrogen atoms (one proton) are known with no neutron, with one neutron and with two

Table 5.1 Atomic weights and valencies

Element	Symbols	Protons	Atomic weight	Electrons	Valency
Hydrogen	H	1	1.0	1	+1
Helium	He	2	4.0	2	0
Carbon	C	6	12.0	2,4	+4
Nitrogen	N	7	14.0	2,5	−3 to +5
Oxygen	O	8	16.0	2,6	−2
Sodium	Na	11	23.0	2,8,1	+1
Magnesium	Mg	12	24.3	2,8,2	+2
Silicon	Si	14	28.0	2,8,4	+4
Sulphur	S	16	32.0	2,8,6	−2 to +6
Chlorine	Cl	17	35.5	2,8,7	−1
Potassium	K	14	39.1	2,8,8,1	+1
Calcium	Ca	20	40.0	2,8,8,2	+2
Bromine	Br	35	80.0	2,8,8,17	−1

Compound	Formula	Ionic weight		Valency
Bicarbonate	HCO_3^-	1 + 12 + 48	= 61	−1
Sulphate	SO_4^{2-}	32 + 64	= 96	−2
Nitrate	NO_3^-	14 + 48	= 62	−1

neutrons. Such varieties of an element's atoms are called *isotopes* (Greek: *isos*, equal and *topos*, place, referring to the place in the Periodic Tables). Hydrogen thus has three isotopes, and they have been given special names and symbols:

    H - common hydrogen, 1 proton
    D - deuterium, heavy stable hydrogen, 1 proton + 1 neutron
    T - tritium, radioactive hydrogen, 1 proton + 2 neutrons

Another way to describe an isotope is to add the number of particles in the nucleus (protons + neutrons) at the upper left corner of the symbol of the element. Thus, the hydrogen isotopes may be written:

    ^1H - common hydrogen, 1 proton
    ^2H - deuterium (also written D), 1 proton + 1 neutron
    ^3H - tritium (also written T), 1 proton + 2 neutrons

Carbon isotopes:

    ^{12}C - common carbon, 6 protons + 6 neutrons
    ^{13}C - heavy stable carbon, 6 protons + 7 neutrons
    ^{14}C - radiocarbon, 6 protons + 8 neutrons

Oxygen isotopes:

    ^{16}O - common oxygen, 8 protons + 8 neutrons
    ^{17}O - heavy (very rare) oxygen, 8 protons + 9 neutrons
    ^{18}O - heavy oxygen, 8 protons + 10 neutrons

Elements, Isotopes, Ions, Units, Errors 59

The relative amounts of the individual isotope species in each element, expressed in percents, are called the *isotopic abundances*. For example, in sea water the relative abundances of hydrogen isotopes are:

^1H : 99.984%
^2H : D: 0:016%
^3H : T ~ 5 × 10^{-6}%

and the isotope abundances of oxygen in sea water are:

^{16}O : 99.76%
^{17}O : 0.04%
^{18}O : 0.20%

The isotopic abundances vary slightly in different natural materials. Tables have been compiled giving the average isotopic abundances as observed in terrestrial materials.

Differences of isotopic compositions observed in groundwaters provide the ground for powerful tracing methods, discussed in chapter 9.

5.3 Atomic weight

The atoms of different elements and of their isotopes differ from each other in their masses or weights. By international convention common oxygen, ^{16}O, has been selected as the basis for a scale of atomic weights: common oxygen, ^{16}O, has been defined as having atomic weight 16. On this basis common hydrogen, ^1H, has an atomic weight of 1, common carbon, ^{12}C, has an atomic weight of 12 and ^{238}U has an atomic weight of 238.

Isotopes have specific atomic weights, according to their number of neutrons. Thus, the atomic weight of ^{35}Cl is 35, that of ^{37}Cl is 37, and the average atomic mass of chlorine is in between: 35.5. The *average atomic weights* of the elements have been determined from two sets of data: the atomic weight of each isotope and the relative abundances of the isotopes. From these data the weighted average atomic weight of each element has been calculated. Atomic weights for elements of hydrochemical interest are included in Table 5.1.

5.4 Ions and valencies

The arrangement of electrons orbiting around atomic nuclei obeys strict rules. As the number of electrons increases, from light atoms to heavy atoms, they are arranged in concentric orbits, constituting shells:

- innermost shell – up to 2 electrons
- second shell – up to 8 electrons
- third shell – again up to 8 electrons
- fourth shell – up to 18 electrons

- fifth shell – again 18 electrons.

In heavier elements the rules are slightly more complicated.

Atoms with complete electron shells are chemically inert. Examples are the noble gases:

He (2 electrons),
Ne (10 electrons, 8 in the outer shell),
Ar (18 electrons, in shells of 2,8,8),
Kr (36 electrons, 2,8,8,18)
Xe (54 electrons, 2,8,8,18,18).

All other atoms tend to interact with each other by sharing electrons of their outer shells. The interacting atoms achieve complete outer shells of 2, 8, or 18 electrons. For example:

Na (electron shells of 2,8,1) + Cl (2,8,7) = NaCl.

In this case the Na gives away its single outer electron, being left with its next complete shell of 8 electrons. As a result, the Na atom loses a negative charge and one of its protons is left non-balanced. In other words, the Na atom in the NaCl atom-combination has one positive, non-balanced, charge. Thus, it is no longer an atom, which was defined as a neutral entity. The new product is called an *ion* and it is marked by a $^+$ sign next to the chemical symbol: Na^+. Similarly, the Cl atom in NaCl gained an electron and completed an outer shell of 8 electrons. Thus, the Cl is now an ion too, but with a negative charge: Cl^-.

Positively charged ions are called *cations* (so called because in electrolysis they move to the cathode) and negatively charged ions are called *anions* (travelling to the anode). Let us examine another example:

$$Ca^{+2} + Cl^- + Cl^- = CaCl_2$$

In this case each of the already familiar chlorine atoms received an electron and was turned into an anion, Cl^-. The calcium atom (20 electrons, arranged in shells of 2,8,8,2) donated its 2 outermost electrons, remaining with a complete outer shell of 8 electrons. Thus, a double charged cation, Ca^{2+}, was formed.

The number of electrons that an atom gives, or gains, is called *valency*. In the examples so far discussed, Cl^- has one (negative) valency and Ca^{2+} has a double (positive) valency. In other words, Ca is bivalent. Valencies of selected elements are given in Table 5.1.

5.5 Ionic compounds

Groups of atoms of certain elements combine in a rather stable state, making up ionic compounds. An example: $CaCO_3$ is the compositional formula of limestone, and from its chemical properties it may be regarded as being composed of $Ca^{2+} + CO_3^{2-}$. A second example: $CaSO_4$ is the compositional formula of anhydrite. From its chemical behaviour it may be regarded as being composed of $Ca^{2+} + SO_4^{2-}$.

In the above examples CO_3^{2-} and SO_4^{2-} are *ionic compounds*. Their ionic weights are the algebraic sum of the atomic weights of the atoms involved: The ionic weight of CO_3^{2-} is $12 + (3 \times 16) = 60$, and the ionic weight of SO_4^{2-} is $32 + (4 \times 16) = 96$. The valencies of ionic compounds are the algebraic sum of the valencies of the atoms involved. Thus, the valency of carbonate is:

$$C^{4+} + O^{2-} + O^{2-} + O^{2-} = CO_3^{2-}$$

and the valency of sulphate is:

$$S^{6+} + O^{2-} + O^{2-} + O^{2-} + O^{2-} = SO_4^{2-}$$

Ionic compounds commonly dissolved in groundwater are: HCO_3^- (bicarbonate), CO_3^{2-} (carbonate), SO_4^{2-} (sulphate), and NO_3^- (nitrate).

5.6 Concentration units

The concentration of ions dissolved in water is expressed in a variety of ways:

Weight per volume. The units most commonly used are *mg/l* for major ions, and *µg/l* (microgram per litre) for trace elements. For example, data for groundwater from the Uriya 4 well are given in Table 5.2 in weight per volume units.

Weight per weight units are commonly applied for highly saline waters. Units in use are, for example, *g/kg* and *mg/kg*. An example, for the Dead Sea brines, is given in Table 5.3. The conversion from weight per volume to weight per weight units is performed by dividing the former by the density of the brine, e.g. 1.23 for the Dead Sea example in Table 5.3.

Ionic equivalence units. The number of anions in solution always equals the number of dissolved cations. Interactions of aqueous solutions with rocks always end up in ionically balanced solutions. Thus, for the discussion of chemical processes it is most meaningful to express chemical data in ionic equivalence units, or in brief, *equivalents*.

$$1 \text{ equivalent} = \frac{\text{number of grams equal to ionic weight}}{\text{valency}}$$

For example, 1 equivalent of Na^+ is $23/1 = 23$ g, and 1 equivalent of SO_4^{2-} is $96/2 = 48$ g.

The data needed to calculate the equivalents of common ions are given in Table 5.1. The equivalent units are too large for convenient use for common

Table 5.2 Dissolved ions, Uriya 4 well, Israel (temperature 28.6°C, pH 7.05)*

	mg/l (ppm)	meq/l
K	7.0	0.18
Na	253	11.0
Ca	84	4.2
Mg	35	2.9
Sr	1.9	
Cl	405	11.4
SO_4	74	1.5
HCO_3	354	5.8
Si	7	
	µg/l (ppb)	
Al	5	
B	510	
Br	90	
Cl	< 0.2	
Cr	4	
Co	3	
Fe	34	
Mn	32	
Ni	5	
P	100	
Total cations		18.3
Total anions		18.7
Total dissolved ions (TDI)		37.0
Reaction error		−1.1%

* From Arad et al. (1984) and Kroitoru (1987). Estimated analytical errors: ±5% for major ions; 10-15% for minor ions.

Table 5.3 Chemical composition of Dead Sea brines (March 1977, density 1.23 g/cm³)*

	K^+	Na^+	Ca^{+2}	Mg^{2+}	Br^-	Cl^-	SO_4^{2-}	TDI
g/l	7.65	40.1	17.2	44.0	5.3	225	0.45	340
g/kg	6.22	32.6	14.0	35.8	4.3	183	0.37	276

* from Kroitoru (1987).

waters and therefore the milliequivalent unit has been introduced, written *meq*. Thus, 1 meq of Na^+ is 23 mg, and 1 meq of SO_4^{2-} is 48 mg.

Milliequivalent units are applied per volume units, e.g. *meq/l*, or per weight units, e.g. *meq/g*.

5.7 Reproducibility, accuracy, resolution and limit of detection

Data should be reported with a description of their quality. For example, the temperature of a spring measured by one person five times in succession was: 16.5°C, 17.2°C, 14.0°C, 15.7°C and 16.9°C. The same spring measured by a second person five times in succession yielded 16.0°C, 16.1°C, 15.8°C, 16.1°C and 16.4°C. A quick glance reveals that the second person's measurements are closer to each other or, to use a more technical expression, the second set of data reveals a higher degree of *reproducibility*. This degree of reproducibility may be expressed quantitatively, by calculating the mean deviation for the two sets of data: The mean value in the first set of temperature measurements is 16.1 and the mean deviation is ± 1.0. Hence, the mean value of the first set of measurements is 16.1 ± 1.0°C. The mean value of the second set of measurements is 16.1 ± 0.2°C. Hence, the reproducibility in the second set was better and the data of this set were of higher quality.

Reproducibility may be best expressed in percentages, e.g. the reproducibility of the first person's temperature measurements was $1.0 \times 100/16.1 = \pm 6.2\%$, whereas the second person's reproducibility was $\pm 1.2\%$.

A second property of data quality is their agreement with standard measures. For example, is a thermometer used in the field, calibrated by comparison to a standard thermometer? Does it show 0°C in a bath of ice and distilled water, and does it show precisely the local boiling temperature? The agreement with standard measures is called *accuracy*. A good thermometer has an accuracy of ± 0.2°C, but many commercial thermometers have a lower accuracy, as they may be off a calibrated instrument by 1.0°C or more. Accuracy of chemical analytical data is tested by the analyst with the aid of solutions that are carefully prepared from chemically pure compounds.

A third property of data quality relates to the *resolution*. Thermometers may be long or short. In the first case readings in 0.2°C intervals are possible, whereas in the second case only 1°C intervals are readable. Thus, the resolution of the long thermometer is 0.2°C and of the short one 1.0°C. Similarly, chemical laboratory data should be reported along with resolution, defined by the instruments involved. Knowledge of the resolution is needed to distinguish differences that are analytically significant. For example, if Cl is determined with a resolution of 0.5 mg/l, then values of 22.7 mg/l and 23.1 mg/l are non-distinguishable, whereas the water of a well with 22.5 mg/l and the water of another well with 25.8 mg/l reveal a difference that is more than twice the resolution, and therefore it is analytically significant (this point is further discussed in section 5.8).

A fourth property of significance in data quality assessment is the *limit of detection*, i.e. the lowest value detectable. This value has to be known in order to process very low concentration values properly. If the limit of detection for Li is 0.015 mg/l, then a value of 0.07 mg/l is analytically significant (this point is further discussed in section 5.8).

5.8 Errors and significant figures

As seen in the previous section, measured values are not absolute, but are obtained with a certain degree of uncertainty. The uncertainty is caused by the combined effect of several error sources. Four major sources for data uncertainty have been described in the previous section: *reproducibility*, *accuracy*, *resolution* and *limit of detection*. To these may be added other factors, e.g. instability of instrumentation, contaminations, accuracy of preparation of standard solutions, etc. The sum of all uncertainties is called the *analytical error*. The analytical error is a cumulative outcome of all errors involved in a measurement. Data included in a laboratory report should always be accompanied by the relevant analytical error, written with a $\pm$ sign, to the right of the result, e.g. 25.62 ± 0.50 mg/l. The analytical error is computed by the various laboratories in slightly different ways, but basically is means that if the same water is analysed 100 times, 67 times the data will be in the given range. Thus, to use the last example, 67% of repeated measurements will fall in the range of 25.62 ± 0.50 mg/l, i.e. 25.12 to 26.12 mg/l.

The analytical error is occasionally expressed as a percentage of the obtained value. Thus 25.62 ± 0.50 mg/l, may also be stated as 25.62 mg/l $\pm 2\%$.

In certain cases the analytical error is not computed for each value, but given in a general mode, e.g. in the bottom line of a table. For example: Analytical errors: Na: ± 0.50 mg/l, Ca: ± 0.70 mg/l, etc., or: Na: $\pm 2\%$, Ca: $\pm 2.5\%$, and so on.

The analytical error is needed to decide which data differ from each other with analytical significance: only data that differ by more than the relevant analytical error should be regarded as different for purposes of data processing. Accordingly, data should be reported only in *significant figures*. SO_4^{2-} concentrations of 16.273 mg/l or 106.16 mg/l are meaningless if the analytical error is, for example, ± 0.7 mg/l. In such a case the data should be reported using only significant figures, namely 16.3 mg/l and 106.2 mg/l.

Reaction error. The sum of cations equals the sum of anions in each solution. Hence, the same should be true for reported laboratory data and the deviation from such an equality provides another way to assess data quality. The equation used is:

$$\text{Reaction error} = \frac{\Sigma_{\text{cations}} - \Sigma_{\text{anions}}}{\Sigma_{\text{ions}}} \times 100$$

The reaction error is thus expressed as a percentage of the total ion concentration. Positive reaction errors indicate cation excess and negative errors indicate anion excess. Reaction errors are caused by:

- The analytical errors of the individual parameters.
- The fact that not all possible ions are commonly measured.

In certain cases it is worthwhile to enlarge the list of ions analysed in order to lower the reaction error: for example, to include No_3^-, Fe^{3+}, or PO_4^{4-}.

At the beginning of each study a decision has to be made which reaction errors will be acceptable. The cut-off at 2% or 5% is common. Analysis with high reaction errors are omitted in the data processing and, if possible, they are discussed with the laboratory personnel.

5.9 Checking the laboratory

Only in rare cases do field hydrochemists themselves measure all the parameters. In most cases samples are sent to laboratories for part, or all, of the measurements. It is the hydrochemist's duty to discuss with the laboratories their data quality, and obtain, at least, the analytical error and limit of detection for each parameter measured. In addition, laboratories should be checked by their clients. There are several kinds of laboratory checks. The most important are:

Duplicate samples. Each batch of samples sent to the lab should include duplicates of one sample, sent with different names and sample numbers. The results for the duplicate sample give a fair picture of the quality of the data. If the duplicates fall in the range of the quoted analytical error, the data for the whole batch of samples is acceptable. If, however, the duplicate values differ by more than the stated analytical error, the results should be discussed with the laboratory personnel and the data of the whole sample batch should be regarded questionable. The differences observed between duplicate samples of several sample batches establish, eventually, the analytical error, of the specific laboratory for each parameter.

Dilution of a sample with measured amounts of distilled water. The results of the diluted sample are acceptable if they agree with the calculated diluted value, within the stated analytical error.

Example: a water sample has been diluted with 1 volume of distilled water. The laboratory results for Mg were 105 mg/l for the non-diluted sample, and 52,9 mg/l for the diluted sample and the analytical error was 0.8 mg/l. Thus, the reported diluted value, 52.9 ± 0.8 mg/l, included in its range the calculated value for 1:1 diluted sample, 52.5 mg/l, and the Mg data of the laboratory may be accepted for the whole batch.

Standard water sample. A highly recommended procedure is to collect a large sample of groundwater, keep it in a cold dark place (to avoid bacterial decomposition) and add a sample of it with every batch of samples sent to the laboratory. This provides a continuous check, revealing the analytical error and serving as a sensitive monitor on the laboratory performance.

5.10 Evaluation of data quality by data processing techniques

Awareness of data quality information is, unfortunately, limited. A large number of papers is published with no analytical errors, or limits of detection, and the results are often stated using non-significant figures. A major source for hydrochemical data are archives with valuable historical data (section 7.5) and these, too, are often without quality descriptions. As a result, investigators tend to discard such data as non-reliable, in spite of their high potential value for the description of early stages of local water exploitation.

Data processing methods provide a substitute for the missing description of analytical data. Repeated measurements of the same water source are occasionally available. In such cases a mean value may be calculated for each parameter (dissolved ions, pH, or temperature). The standard deviations of these values serve as an estimate of the analytical error. In fact, these standard deviations also incorporate the natural fluctuations in the measured water source. Therefore, these mean deviations serve as conservative estimates to the analytical errors. These estimated errors may then be applied to all the data reported from the same laboratory during the same period. Experience shows that old data are often good and acceptable.

Reproducibility deduced from clustering of data. A basic step in the processing of hydrochemical data is to plot the various parameters as a function of the total dissolved ions (TDI) or other parameters (sections 6.3–6.4). Occasionally, a group of hydrologically related samples cluster, or group around the same values (e.g. wells tapping the same aquifer). In such cases the mean deviation from the mean value, observed for each dissolved ion, serves as an upper limit for the analytical error (it includes also the natural fluctuations).

Reproducibility deduced from data plotting on a mixing, or dilution line. As discussed in section 6.6, data occasionally plot along a well-defined line (in most cases a mixing line). The mean deviation from such a line provides an upper limit for the sum of analytical errors of the parameters plotted.

5.11 Putting life into a dry table

Mastering the units and data quality concepts discussed in the present chapter is absolutely necessary in order to proceed with the following chapters of this book, and to enjoy hydrochemistry.

Elements, Isotopes, Ions, Units, Errors

A new hydrochemical study, commenced in Wonderland, included the collection of *five identical* sample bottles from the Wisdom Spring. They were transferred to a laboratory at different dates for the determination of major dissolved ions and the data are given in Table 5.4. What can be deduced, or calculated, in light of the data quality concepts and data processing approaches, discussed in the previous sections?

Table 5.4 Dissolved ions, Wisdom Spring, meq/l*

Sample No.	K^+	Na^+	Ca^{2+}	Mg^{2+}	Cl^-	HCO_3^-	SO_4^{2+}
1	0.40	4.52	1.23	0.91	43.2	1.95	0.21
2	0.37	4.93	1.44	0.97	4.75	2.46	0.01
3	0.41	5.24	1.67	1.03	5.11	2.72	0.17
4	0.37	4.92	1.35	1.09	5.01	2.51	0.02
5	0.33	4.67	1.23	0.95	4.82	2.03	0.1

* Perfect Analytics Laboratory, Chemistryland.

Table 5.5 Mean concentrations (meq/l), mean deviations and reproducibilities for the Wisdom Spring data (Table 5.4)

Sample	K^+	deviation	Na^+	deviation	Ca^{2+}	deviation	Mg^{2+}	deviation
1	0.40	+0.02	4.52	−0.34	1.23	−0.15	0.91	−0.08
2	0.37	+0.01	4.93	+0.07	1.44	+0.06	0.97	−0.02
3	0.41	+0.03	5.24	+0.48	1.67	+0.29	1.03	+0.04
4	0.37	−0.01	4.92	+0.06	1.35	−0.03	1.09	+0.10
5	0.33	−0.05	4.67	−0.19	1.23	−0.15	0.95	−0.04
Mean	0.38	±0.03	4.86	±0.23	1.38	±0.14	0.99	±0.06
Reproducibility		±7.9%		±4.7%		±10%		±6.1%

Sample	Cl^-	deviation	HCO_3^-	deviation	SO_4^{2-}	deviation
1	4.32	−0.48	1.95	+0.38	0.21	+0.11
2	4.75	−0.05	2.46	+0.13	0.01	=0.09
3	5.11	+0.31	2.72	+0.39	0.17	+0.07
4	5.01	+0.21	2.51	+0.18	0.02	−0.08
5	4.82	+0.02	2.03	−0.30	0.11	+0.01
Mean	4.80	±0.21	2.33	±0.28	0.10	±0.07
Reproducibility		±4.5%		±12%		±70%

Sending several bottles of the same water collection provides the data needed to calculate the laboratory's reproducibility (section 5.7). This procedure is recommended at the beginning of a hydrochemical study, involving a new laboratory, or as a periodical check of a known laboratory (section 5.9).

From the data of Table 5.4 it can be seen that the Cl value of sample 1 is significantly higher than the Cl values of the other four samples. An inquiry

Table 5.6 Total ions (meq/l) and reaction errors, calculated for the Wisdom Spring data (Table 5.4)

Sample	K^+	Na^+	CA^{2+}	Mg^{2+}	Cl^-	HCO_3^-	SO_4^{2-}	Total cations	Total anions	Total ions	Reaction error
1	0.40	4.52	1.23	0.91	4.32	1.95	0.21	7.06	6.48	13.54	4.3%
2	0.37	4.93	1.44	0.97	4.75	2.46	0.01	7.71	7.22	14.93	3.3%
3	0.41	5.24	1.67	1.03	5.11	2.72	0.17	8.35	8.00	16.35	2.1%
4	0.37	4.92	1.35	1.09	5.01	2.51	0.02	7.73	7.54	15.27	1.2%
5	0.33	4.67	1.23	0.95	4.82	2.03	0.11	7.18	6.96	14.14	1.6%

Table 5.7 Dissolved ions in samples of Wisdom Spring (mg/l)

Sample	K^+	Na^+	Ca^{2+}	Mg^{2+}	Cl^-	HCO_3^-	SO_4^{2-}	TDI
1	15.6	104	24.6	11.0	153	119	10.1	428
2	14.5	113	28.8	11.8	169	150	0.5	488
3	16.0	120	33.4	12.5	181	166	8.2	537
4	14.5	113	27.0	13.2	178	153	1.0	500
5	12.9	107	24.6	11.5	171	124	5.3	456

to the laboratory may reveal that it was a typing error, and 43.2 should be corrected to 4.32. The mean Na value is:

$$\frac{4.52 + 4.93 + 5.24 + 4.92 + 4.67}{5} = 4.856$$

or, in *significant figures*, 4.86 meq/l.

The deviation of each Na measurement from the average is given below:

Sample	Na	Deviation
1	4.52	−0.34
2	4.93	+0.07
3	5.24	+0.48
4	4.92	+0.06
5	4.67	−0.19
mean	4.86	0.23

Thus, the value of 4.52 reported for the Na in sample 1, deviates from the mean value by:

$$4.86 - 4.52 = -0.34 \text{ meq/l}$$

A minus sign is added to specify that measurement No. 1 was lower than the mean. The sum of negative deviations (in our example: −0.34 and −0.19, sum −0.53) should be close to the sum of the positive deviations (in our case 0.07 + 0.48 + 0.06 = 0.61), indicating that the mean value (4.86) and the deviations have been correctly calculated. The two sums may slightly differ (0.53 versus 0.61 in the example) due to rounding up of each value to its significant figures.

The *mean deviation* is calculated from all deviations regardless of their positive or negative sign. In the present example:

$$\frac{0.34 + 0.07 + 0.48 + 0.06 + 0.19}{5} = 0.228$$

or, in significant figures, **0.23**.

The mean deviation of the Na measurements reported in Table 5.4 is, thus, ± 0.23 and the mean Na concentration is: 4.86 ± 0.23 meq/l.

The mean deviation, or the *standard deviation* (more familiar to some students), are numerically very close. They are often also called *sigma* (after the Greek letter). Data are commonly reported with one sigma (4.86 ± 0.23 in the present example of Na data). Statistically, if additional measurements of the same type will be done, *67% of the cases will fall in the one-sigma range*, 90% of the cases will fall in the two-sigma range (4.86 ± 0.46 in our Na example), and 97% of the cases will fall in the three-sigma range. By convention, results are reported with *one sigma*, unless otherwise specified. Table 5.4 has been reworked in Table 5.5, which includes also the average, or mean, concentrations of the various ions, and the reproducibilities.

The total concentration of dissolved cations, in short: *total cations*, is the sum of the concentrations of K, Na, Ca and Mg, i.e.:

$0.401 + 4.52 + 1.23 + 0.91 =$ **7.06 meq/l**

The value of *total anions* is:

$4.32 + 1.95 + 0.21 =$ **6.48 meq/l**

The value of total dissolved ions (TDI) is:

$7.06 + 6.48 = 13.54$ meq/l

The *reaction error* (section 5.8) is the difference of total cations and total anions, expressed as percentages of the TDI. In the example of sample 1. Table 5.4:

$$\frac{(7.06 - 6.48)}{13.54} \times 100 = 4.3\%$$

The data of Table 5.4 have been reworked again in Table 5.6, which includes the total cations, total anions, total ions, and reaction errors. The data in Table 5.4 have been expressed in meq/l. However, most people are used to the mg/l units for assessment of the degree of salinity of water. The conversion of meq/l data into mg/l has been discussed in section 5.6, and is here demonstrated on the Wisdom Spring data (Table 5.7):

Na: $\dfrac{4.52 \times 23.0}{1} = 104$ mg/l

K^-: $\dfrac{0.40 \times 39.1}{1} = 15.6$ mg/l

Ca^{2+}: $\dfrac{1.23 \times 40.0}{2} = 24.6$ mg/l

Mg^{2+}: $\dfrac{0.91 \times 24.3}{2} = 11.0$ mg/l

Cl^-: $\dfrac{4.32 \times 35.5}{1} = 153$ mg/l

HCO_3^-: $\dfrac{1.95 \times 61}{1} = 119$ mg/l

SO_4^{2-}: $\dfrac{0.21 \times 96}{2} = 10.1$ mg/l

Comparing the raw data in Table 5.4 with Tables 5.5, 5.6 and 5.7 and the relevant discussion, the reader may, perhaps, be amazed how much life could be put into a single table of dry data.

5.12 Evaluation of calculated reproducibilities and reaction errors

Reproducibility of measurements of major dissolved ions can, in theory, be better than $\pm 1\%$. However, in real life poorer reproducibilities are common. In Table 5.5 the reproducibility values of $\pm 4.7\%$ of the Na data and $\pm 4.5\%$ for the Cl data are on the limit of acceptance. However, $\pm 10\%$ for Ca and $\pm 12\%$ for HCO_3 are shaky and the value of $\pm 70\%$ of SO_2 is totally non-acceptable.

In the Wisdom Spring example, the investigator may discuss the results with the laboratory staff and see whether they can improve. Otherwise, it is necessary to shift to another laboratory and repeat the check.

As stated, reaction errors are another way to estimate data quality (section 5.8). The reaction error can (and should) be better than 1%, but they are often larger. Many investigators will accept up to 5%. Using the latter criterion the reaction errors calculated in Table 5.6 (4.3%, 3.3%, 2.1%, 1.2%, and 1.6%) are acceptable. This example brings out a major shortcoming of the reaction error quality test: the poor reproducibility of the SO_4 data, demonstrated above, is not reflected in the reaction error because SO_4 is a minor compound of the Wisdom Spring water. Low reaction errors indicate acceptable quality of the data of the major ions alone.

The reaction errors in repeated measurements are expected to be random, i.e. in part of the measurements the total cations will exceed the total anions and in the rest of the cases the total cations will be less than the total anions. However, occasionally a sytematic pattern is seen. In Table 5.6 all the measurements reveal total cations to be higher than the total anions (check it). Possible explanations:

- A systematic analytical error on one (or more) of the ions.
- The list of measured anions should include another anion that happens to be important, e.g. nitrate (CO_3^-) or bromide (Br^-).

6 CHEMICAL PARAMETERS – DATA PROCESSING

6.1 Data tables

Hydrochemical studies generate large amounts of data describing different parameters, obtained in the field and reported by various laboratories. The first stage in data processing is to organize the data into tables. This stage is important and warrants some thinking. Have a look at Tables 6.1, 6.2 and 6.3. They contain the same data, but differ in their structure. Which of the three tables is 'impossible', and which is most handy and most informative?

Table 6.1 has no caption that relates the data to specific wells or springs; the analytical units are not specified; and the data are arranged in increasing order of sample number, which has no meaning. Table 6.1 is useless.

Table 6.2 has a caption relating the data to the Mice Springs: it gives the names of the springs from which the samples were collected, the units (meq/l), the name of the laboratory that determined the dissolved ions, the sum of dissolved ions (TDI), and the range of analytical errors. The columns are arranged with the field data first (on the left), followed by the cations and finally (on the right) the anions. The lines of the table are arranged in order of increasing concentration of TDI. A glance at this table reveals that the temperature and the concentrations of the various ions covary, i.e. these parameters are positively correlated.

Table 6.3 is even more organized: the cations are arranged in order of increasing concentration, $K < Mg < Ca < Na$ and the anions are arranged by increasing concentrations in the most saline sample (spring H), namely $SO_4 < HCO_3 < Cl$. This order is seen in all the samples, except the first (spring A). So Table 6.3 reveals that we are dealing with a complex of springs that have similar *relative* abundances of the dissolved ions, but differ in salinity (TDI) and temperature.

There is no one correct way to organize tables – the optimal solution has to be sought out in each case. Tables should not be overloaded: occasionally, splitting of data into several tables is beneficial.

Chemical Parameters – Data Processing

Table 6.1

Sample No.	K	Cl	Mg	Na	SO$_4$	Ca	HCO$_3$	Temp. (°C)
71	0.60	9.11	1.78	8.10	1.09	2.72	3.00	21.4
72	1.03	15.5	2.76	11.5	1.36	3.84	4.22	26.1
73	0.02	0.53	0.47	0.91	0.73	1.22	1.36	15.2
74	0.16	2.67	0.80	2.71	0.82	1.60	1.76	16.8
75	0.31	4.82	1.13	4.51	0.91	1.97	2.17	18.3
76	0.75	11.3	2.10	9.90	1.18	3.10	3.41	23.0
77	1.18	17.7	3.08	15.3	1.45	4.21	4.62	27.6
78	0.89	13.3	2.43	11.7	1.27	3.47	3.81	24.5
79	0.46	6.97	1.46	6.30	1.00	2.35	2.59	19.9

Table 6.2 Chemical composition of the Green Mice Springs complex (meq/l)*

Sample No.	Spring	Temp. (°C)	K	Na	Ca	Mg	Cl	HCO$_3$	SO$_4$	TDI
73	A	15.2	0.02	0.91	1.22	0.47	0.53	1.36	0.73	5.24
74	C	16.8	0.16	2.71	1.60	0.80	2.67	1.76	0.82	10.5
75	E	18.3	0.31	4.51	1.97	1.13	4.82	2.17	0.91	15.8
79	B	19.9	0.46	6.30	2.35	1.46	6.97	2.59	1.00	21.1
71	D	21.4	0.60	8.10	2.72	1.78	9.11	3.00	1.09	26.4
76	I	23.0	0.75	9.90	3.10	2.10	11.3	3.41	1.18	31.7
78	F	24.5	0.89	11.7	3.47	2.43	13.4	3.81	1.27	37.0
72	G	26.1	1.03	13.5	3.84	2.76	15.5	4.22	1.36	38.8
77	H	27.6	1.18	15.3	4.21	3.08	17.7	4.62	1.45	47.5

* Big Chemistry Laboratory, Dataland. Temperature measurement error: ± 0.2°C; analytical errors are ± 2% for Na, Ca, Cl and HCO$_3$; and ± 5% for K, Mg and SO$_4$.

Table 6.3 Rearranged chemical composition data of the Green Mice Springs complex (meq/l)*

Sample	Spring	Temp.	K	Mg	Ca	Na	SO$_4$	HCO$_3$	Cl	TDI
73	A	15.2	0.02	0.47	1.22	0.91	0.73	1.36	0.53	5.24
74	C	16.8	0.14	0.80	1.60	2.71	0.82	1.76	2.67	10.5
75	E	18.3	0.31	1.13	1.97	4.51	0.91	2.17	4.82	15.8
79	B	19.9	0.46	1.46	2.35	6.30	1.00	2.59	6.97	21.1
71	D	21.4	0.60	1.78	2.72	8.10	1.09	3.00	9.11	26.4
76	I	23.0	0.75	2.10	3.10	9.90	1.18	3.41	11.3	31.7
78	F	24.5	0.89	2.43	3.47	11.7	1.27	3.81	13.4	37.0
72	G	26.1	1.03	2.76	3.84	13.5	1.36	4.22	15.5	38.8
77	H	27.6	1.18	3.08	4.21	13.3	1.45	4.62	17.7	47.5

* Big Chemistry Laboratory, Dataland. Temperature measurement error: ± 0.2°C; analytical errors are ± 2% for Na, Ca, Cl and HCO$_3$; and ± 5% for K, Mg and SO$_4$.

6.2 Fingerprint diagrams

Figure 6.1 contains the data of dissolved ions given in Table 6.3. In this figure each spring is represented by one line that provides a visual description of:

- The relative concentrations (the position of the line at the upper or lower part of the diagram.
- The relative salinities (the position of the line at the upper or lower part of the diagram).

Thus, each line is the compositional imprint of a water sample and various samples may be compared to each other in the way that people may be classified and identified by their fingerprints. Historically, it was Schoeller who in 1954 applied a fingerprint diagram for the first time in relation to groundwater analyses. The fingerprint diagram is a most powerful tool in the hands of the hydrochemist, provided that it is well prepared with regard to the following points.

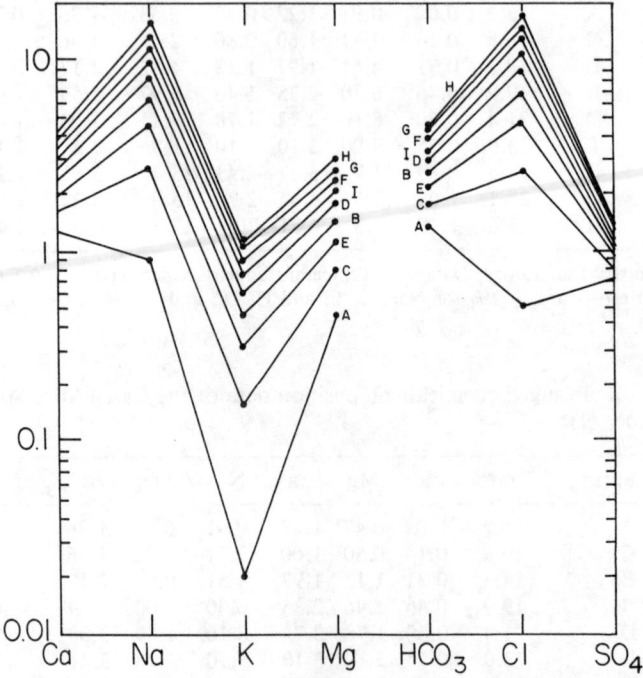

Fig. 6.1 A fingerprint diagram of the data of Table 6.3. Cations are, by convention, plotted on the left and anions on the right. In Fig. 6.5 the same data have been re-plotted in increasing order of cation concentration and decreasing order of anion concentration.

Chemical Parameters – Data Processing

The logarithmic concentration axis. The set of data given in Table 6.4 has been generated so that the concentration of dissolved ions increases from data set 1 to set 5, but the relative abundances of the ions are preserved (this can be checked, for example by comparing the Mg:Ca ratio in data sets 1 to 5). This imitates dilution of a saline water by different amounts of a fresh (ideally, 'distilled') water, a common occurrence in nature. The data of Table 6.4 have been plotted once on regular millimetre graph paper with a *linear* concentration axis (Fig. 6.2), and once on *semi-logarithmic* graph paper, with a logarithmic concentration axis (Fig. 6.3). The outcome is striking: the same data, plotted with a different concentration axis, reveal intrinsically different patterns. On the regular millimeter paper the lines of the individual water samples *differ in their angles*, whereas on the semi-logarithmic paper the lines have the *same angles*. The latter pattern reflects the dilution of a saline water with fresh water, whereas the former diagram is misleading in giving the impression that each plotted water sample has its own relative ion abundances. For this reason semilogarithmic paper should always be used to produce fingerprint diagrams.

Another advantage of semilogarithmic paper is that it provides equal room for ions of low concentrations as for ions of high concentration: the distance between the K points of samples 1 and 5 on Fig. 6.3 is the same as the distance between the Na points of samples 1 to 5. Not so in the millimetre paper diagram (Fig. 6.2), in which the K and SO_4 points are squeezed and the Na and Cl points are overspread.

Table 6.4 Synthetic data generated to simulate dilution of a saline water by a fresh water (meq/l)

Data set	K	Mg	Ca	Na	SO_4	HCO_3	Cl	TDI
1	0.2	0.8	1.4	2.6	0.3	2.4	2.3	10.0
2	0.4	1.6	2.8	5.2	0.6	4.8	4.6	20.0
3	0.6	2.4	4.2	7.8	0.9	7.2	6.9	30.0
4	0.8	3.2	5.6	10.4	1.2	9.6	9.2	40.0
5	1.0	4.0	7.0	13.0	1.5	12.0	11.5	50.0

Selecting the right number of cycles of the semilogarithmic scale. Several computer programs are available to facilitate the preparation of fingerprint diagrams, but in order to master them it is necessary to be familiar with the principles. The number of cycles needed is determined by the lowest and highest concentration values in the data set. For example, in Table 6.3 the lowest concentration value is the K concentration in spring A(0.02 meq/l) and the highest is the concentration of Cl in spring H(17.7 meq/l). Thus, four logarithmic cycles were needed to plot the data of Table 6.3 in the fingerprint diagram of Fig. 6.1 (check it). Examples of different semilogarithmic papers are given in Fig. 6.4.

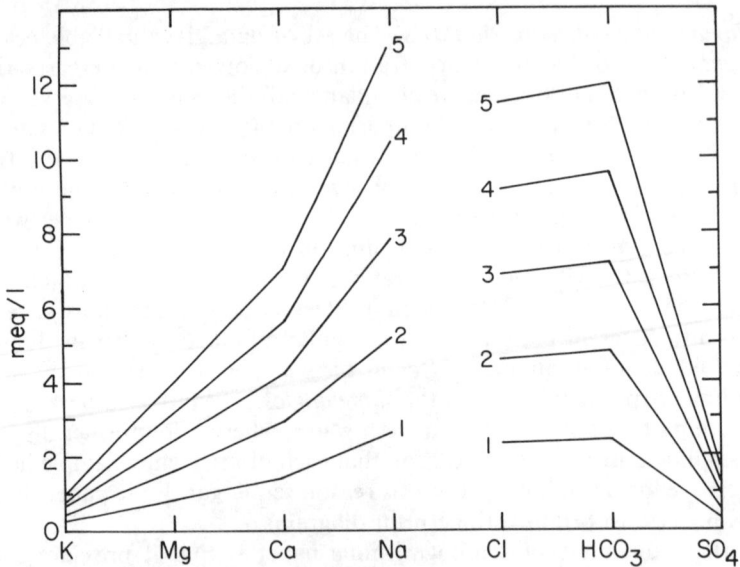

Fig. 6.2 A linear fingerprint diagram of samples formed by different degress of dilution of a saline water (Table 6.4). The data reveal compositional lines of different patterns, although their relative ion abundances are the same.

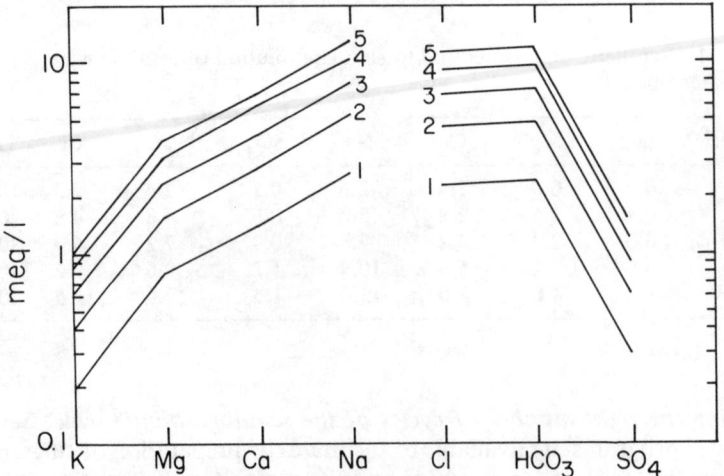

Fig. 6.3 The data of Fig. 6.2 (Table 6.4) re-plotted on a semilogarithmic paper. Parallel lines of the same pattern are obtained, reflecting different degrees of dilution of a saline water.

Selection of the order of ions on the horizontal axis. Figures 6.1 and 6.5 portray the data of Table 6.3, the difference being the order of the ions on the horizontal axis: in the first case the order is Ca, Na, K, Mg, HCO$_3$, Cl,

Chemical Parameters – Data Processing

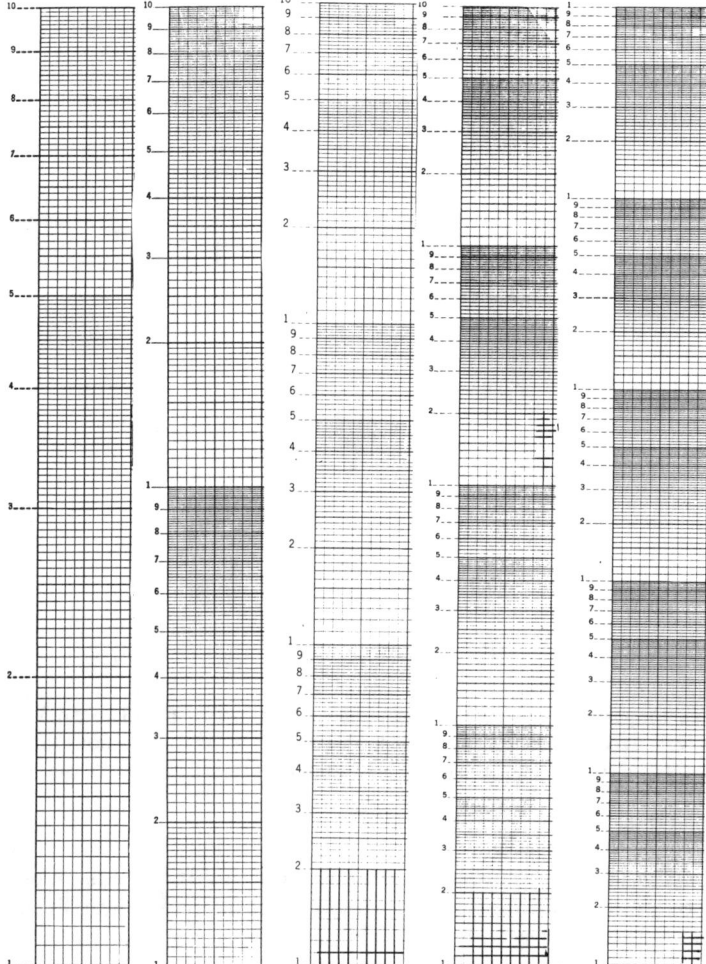

Fig. 6.4 Semilogarithmic papers with one, two, three, four and five cycles. The five-cycle paper is most convenient for the majority of hydrochemical data processing tasks.

SO_4, whereas in the second case the order is K, Mg, Ca, Na, Cl, HCO_3, SO_4. The resulting visual images are completely different. Two considerations influence the decision on the order of ions in the fingerprint diagram:

- placing ions of geochemical importance close to each other, e.g. K close to Na or
- arranging the ions by concentration, as has been done in Fig. 6.5.

The advantage of this mode is that simple chemical imprint lines are obtained, so that pattern differences between the lines on the diagram are more obvious. Zigzag lines, as in Fig. 6.1, are more confusing.

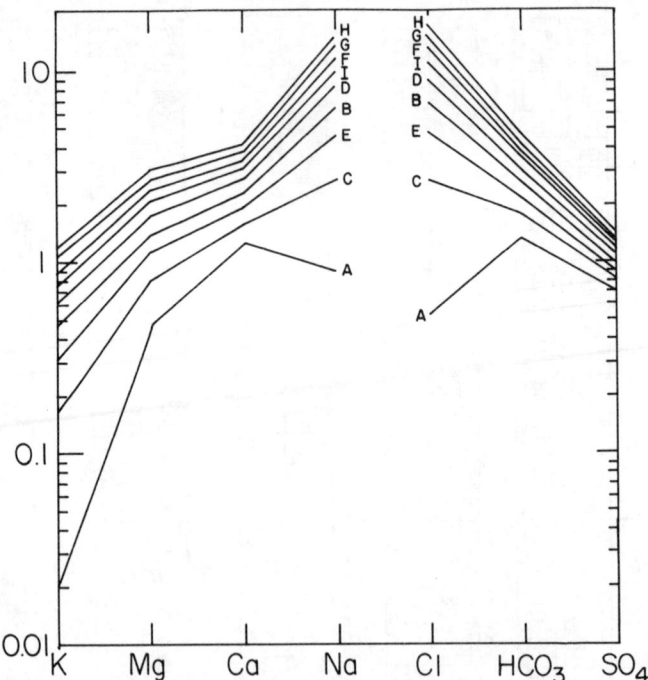

Fig. 6.5 Fingerprint diagram of the same data as in Fig. 6.1 (Table 6.3), but with cations arranged in increasing order of concentration and anions arranged in decreasing order of concentration, resulting in simple lines that can readily be compared.

Another consideration is demonstrated in Fig. 6.5: the cations are arranged in increasing order of concentration and the anions in decreasing order of concentration. In this way the cation and anion lines acquire symmetry and are easy to follow (imagine Fig. 6.5 with the anions arranged in increasing order of concentrations - what would be the outcome?).

Separation of cations and anions. As a convention it is suggested that the cations should be plotted to the left and the anions to the right of the fingerprint diagram. This separation is in accordance with geochemical thinking: waters are described by the concentration order of the cations and the anions (section 6.9) and water-rock interactions are discussed by cations and balancing anions (section 6.8). For clarity it is recommended to have two line segments for each water sample - a line connecting all cations and a separate line connecting all anions, as done in Figs. 6.3 and 6.5.

A methodological note. The method of construction of fingerprint diagrams discussed here has been found most informative by the present author. Other combinations of the fingerprint diagram principles are currently used by various investigators. As a result, different types of fingerprint diagrams are included in the discussion of case studies in the following chapters.

6.3 Composition diagrams

Pairs of measured parameters may be plotted in x-y (scatter) diagrams, or composition diagrams. These may be closely placed on one page, as shown in Fig. 6.6 for the data given in Table 6.3. The composition diagram of Fig. 6.6 portrays the composition of the Green Mice Springs in a visual form. The following features can be observed:

- The springs vary considerably in their concentrations.
- The data plot on straight lines, revealing a positive correlation of K, Na, Ca, Mg, Cl, HCO_3 and SO_4 with TDI.

Similarly, the temperature is seen to be positively correlated with the TDI. The composition diagram provides a handy means of visual expression of large amounts of data, complementing the fingerprint diagram, each of the two having its advantages.

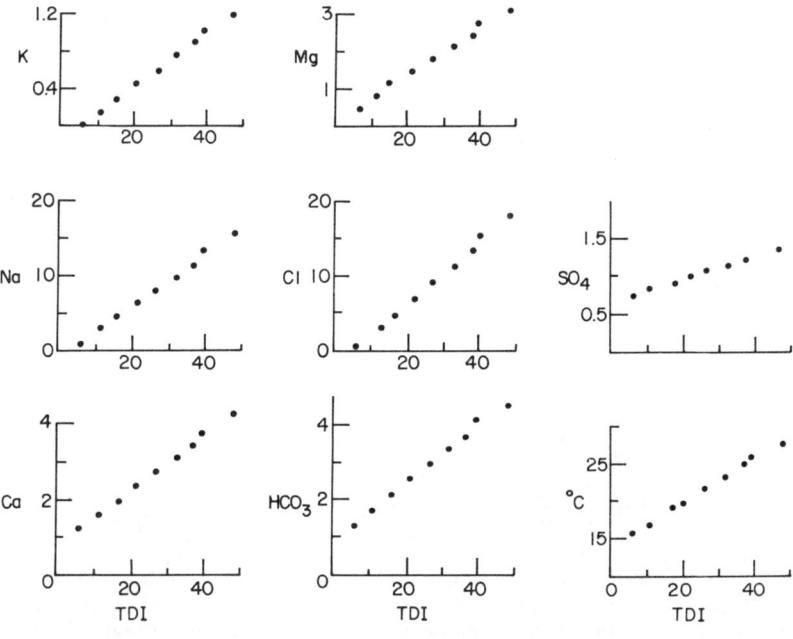

Fig. 6.6 A composition diagram of the Green Mice Springs (Table 6.3). Concentrations are given in meq/l. The compositional interrelations of the nine springs are clearly exhibited: mixing lines indicate a saline warm end member mixes in various proportions with a fresh cold end member.

6.4 Major patterns seen in composition diagrams

Let us plot Cl against TDI for data obtained for samples collected in regional studies carried out on different water sources. The following patterns are possible:

A cluster (Fig. 6.7). Cl concentrations in five adjacent springs reveal a cluster when plotted against TDI. This may indicate that all five springs are fed by one type of water, or, in hydrological terms, the five springs are fed by the same aquifer. One may argue that different waters are involved, all having the same Cl concentration. Therefore, other parameters are plotted as well, Ca and SO_2 in the example of Fig. 6.7. If the same pattern of a single cluster is obtained, the conclusion of one type of water is confirmed.

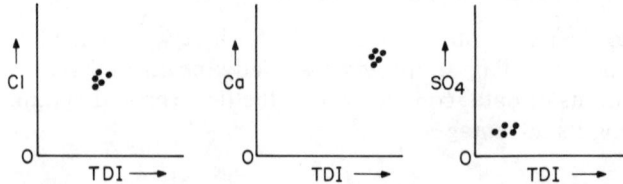

Fig. 6.7 A cluster pattern in a set of composition diagrams of five adjacent springs. The pattern indicates one type of water is involved. The more parameters are checked, the higher is the confidence of the conclusion.

Two clusters (Fig. 6.8). Two clusters of springs are revealed in plots of Mg, Na and temperature versus chloride levels. Two water groups emerge: a group of five springs is fed by a water that has relatively higher Mg, Na and Cl concentrations and is relatively cold; and a group of three springs is fed by water with relatively lower Mg, Na and Cl concentrations and a higher temperature.

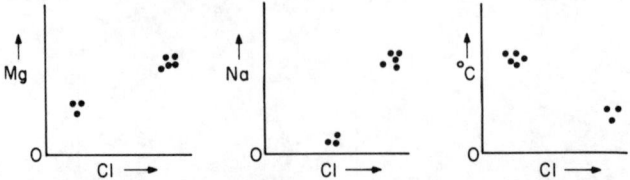

Fig. 6.8 Two clusters in a set of eight adjacent springs. The pattern indicates two distinct types of water occur in the studied region (with no intermixing): a water type of low Cl, Mg, Na and elevated temperature, and a water type of high Cl, Mg and Na and a low temperature.

Data plotting on lines is shown in Figs. 6.9a-c. Such lines may be of different kinds:

Chemical Parameters – Data Processing

- Figure 6.9a depicts a line *extrapolating to the zero points*.
- Figure 6.9b shows a line *extrapolating to a value on the TDI axis*.
- Figure 6.9c *extrapolates to a value on the vertical axis*.

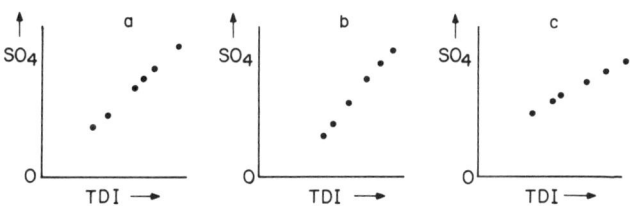

Fig. 6.9 Data from well fields plotting in straight lines in composition diagrams. These are mixing lines, of which three variations are shown: (a) the line extrapolates to the zero points, indicating mixing of a saline water with a water that has negligible SO_4 concentrations (dilution); (b) the line extrapolates to a point in the TDI axis, indicating the fresher end member contains significant concentrations of ions other than SO_4; and (c) the line extrapolates to the SO_4 axis, indicating both intermixing waters contain significant concentrations of SO_4.

Such lines are formed by the mixing of fresh and saline water in various percentages, e.g. water ascending from depth and intermixing with seasonal rain and snowmelt recharge. The topic is further discussed in sections 6.6 and 6.7.

Triangular distribution. Occasionally data plot in triangles in composition diagrams, as seen in the examples given in Fig. 6.10. Such a distribution points to intermixing of three distinct water types, as seen in Fig. 6.10:

- A water type of low TDI, high Na and low Mg and Cl.
- A water type of medium TDI, low Na and high Mg and Cl.
- A water type of high TDI, high Na, low Mg and medium Cl.

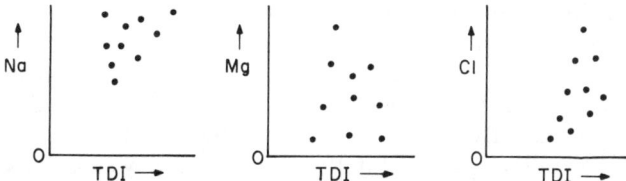

Fig. 6.10 Data of a farm well falling in triangular areas on compositional diagrams, indicating three distinct water types intermix in varying proportions.

Random distribution. Figure 6.11a shows a case where data from a studied region reveal a random distribution on a composition diagram. Random distributions of data may indicate:

- The measured samples are of non related water sources of different compositions.

• The analytical quality of the data is poor.

The latter case may be established if other pairs of parameters show a distinct pattern. In the example of Fig. 6.11 the SO_4 reveals a random distribution as a function of TDI, but Cl and HCO_3 reveal distinct patterns of mixing between fresh and saline end members. Thus, the SO_4 values are suspected to be erroneous and this parameter has to be remeasured, to check for a distinct pattern.

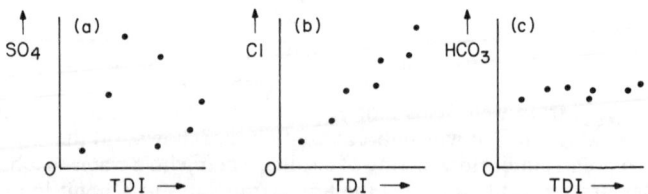

Fig. 6.11 Compositional diagrams of a set of well samples. (a) random SO_4 distribution; (b) a positive Cl - TDI correlation; and (c) a constant HCO_3 value. Possible interpretation: mixing of fresh and saline end members that both have the same HCO_3 concentration; the SO_4 measurements are suspected to be erroneous and should be repeated.

6.5 Establishing hydraulic interconnections

Direct proofs for hydraulic interconnections

If dyes injected into one well are then found in an adjacent well, this directly proves hydraulic interconnections and direction of groundwater flow. Fungal spores, salt, and various radioactive isotopes are other tracers that have been used to trace groundwater flow.

A drop of the water table as a result of pumping in an adjacent well is another direct proof for hydraulic interconnections.

The drawback of these methods is that they can be traced over small distances, of the order of tens to thousands of metres, and even that requires substantial efforts. For greater distances, other indirect tracing methods are at our disposal. But first let us see why it is important to establish hydraulic interconnections.

Water level gradients – a necessary condition in determining flow direction, but not a sufficient one

Water flows from high to low points, and hence a water level gradient is an essential condition for underground flow. A common practice among hydrologists is to reconstruct groundwater flow directions based on water level gradients (section 4.3). However, a glance at Figs. 6.12–6.14 reveals that a gradient is a necessary but insufficient condition. Figure 6.12 portrays three wells tapping the same aquifer and the water flows from the region of well I to well II and on to well III, down gradient. In Fig. 6.13 three wells manifest a relatively high water table at well I, a medium water table at well II and a relatively low water table at well III. However, the three wells are not interconnected – they are separated by aquicludes. The three wells shown in Fig. 6.14 are separated by a buried anticline, disconnecting well I from wells II and III, in spite of the apparent water table gradient. Wells II and III are hydraulically interconnected.

As already mentioned in section 4.4, one can never deduce flow directions from water levels alone.

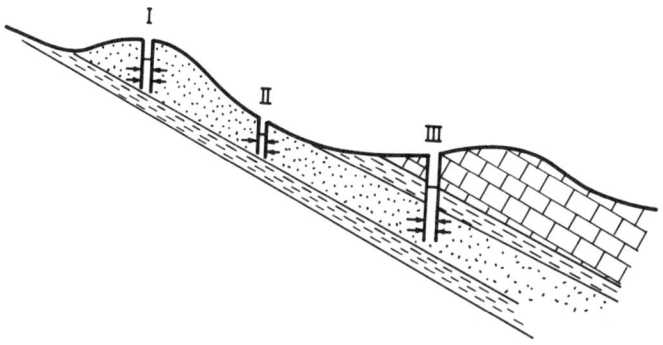

Fig. 6.12 Three wells tapping the same aquifer. Water flows down gradient from the area of well I to well II and on to well III.

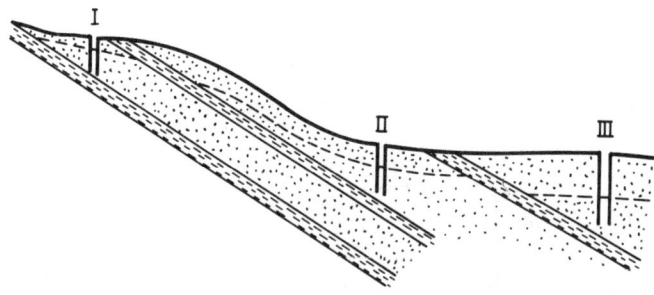

Fig. 6.13 Three wells with water tables similar to those seen in Fig. 6.12, but penetrating distinct aquifers that are separated by aquicludes. They have no hydrological connections, in spite of the *apparent* water level gradient.

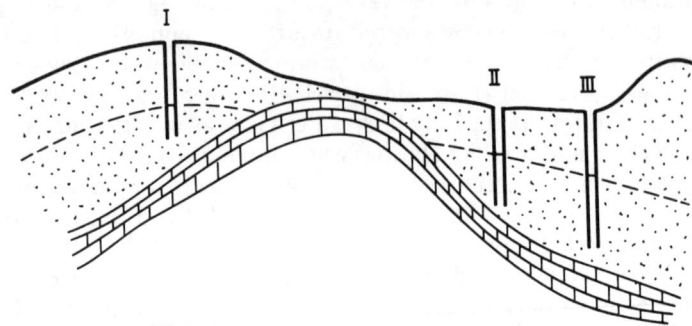

Fig. 6.14 Three wells with an apparent water table gradient of well 1 > well II > well III. However, a concealed folded structure isolates well I from wells II and III.

The use of groundwater parameters to establish hydraulic interconnections

The patterns of data points distribution on composition diagrams, discussed in section 6.4, provide the means of checking hydraulic interconnections:

Clustering around a single value, as in Fig. 6.7, indicates that the sampled springs belong to the same type of water and are, therefore, most likely interconnected.

Several clusters, as in Fig. 6.8, indicate that separate hydraulic systems are involved, having distinct water types and isolated from each other.

Data plotting on lines (Fig. 6.9) are likely to indicate mixings of two water types in various proportions, a topic further discussed in the next section. Mixing of two water types indicates that at some point two distinct water systems are locally interconnected, either naturally or by manmade drillings. The same holds true for triangular data patterns on composition diagrams, indicating that three distinct water types are interconnected at some place.

A practical example

The data of Table 6.5 have been plotted on a fingerprint diagram in Fig. 6.15. Three water groups emerge, called A, B, and C. The groups are homogeneous, i.e. the samples of each group have nearly the same ion concentrations. Figure 6.16 depicts the data of Table 6.5 on composition diagrams. The three distinct composition groups are clearly visible. Now that a compositional picture has been gained, Table 6.5 may be reorganized by water groups and by increasing concentration of cations and anions. The

Chemical Parameters - Data Processing   85

outcome, Table 6.6, is informative and reflects the composition pattern of the Hot Fudge wells – they are not fed by one uniform aquifer, but tap three distinct groundwater systems. The wells of group A are most likely hydraulically interconnected, and so are the wells of group B and of group C. But the water systems A, B, and C are hydraulically not connected.

The next stage should deal with a search for a geological, hydrological, or geographical meaning of the observed composition groups, turning them into geochemical groups.

Table 6.5 Dissolved ions in the Hot Fudge well field (meq/l)*

No.	Na	Cl	SO_4	Mg	HCO_3	Ca	K
1	7.2	8.1	0.52	1.5	3.8	3.7	<0.02
2	3.8	4.5	3.1	2.2	5.6	6.0	1.2
3	4.5	5.2	3.7	2.7	5.1	5.5	1.0
4	11.5	11.0	5.4	4.0	9.5	7.2	2.7
5	6.9	8.2	0.84	1.2	3.1	4.0	0.04
6	11.0	12.0	6.0	4.6	8.8	8.3	2.7
7	7.5	8.6	0.60	1.8	4.2	4.1	<0.02
8	12.0	11.5	5.8	4.4	8.3	7.8	2.4

* Good Day Laboratory; analytical errors: ± 4% for K, Mg, Ca, SO_4 and HCO_3.

Table 6.6 Reorganized table of the Hot Fudge data (Table 6.5) (meq/l)

Group	No.	K	Mg	Ca	Na	SO_4	HCO_3	Cl	TDI
A	1	<0.02	1.5	3.7	7.2	0.52	3.8	8.1	24.8
	7	<0.02	1.8	4.1	7.5	0.60	4.2	8.6	26.8
	5	0.04	1.2	4.0	6.9	0.84	3.1	8.2	24.3
B	2	1.2	2.2	6.0	3.8	3.1	5.6	4.5	26.4
	3	1.0	2.7	5.5	4.5	3.7	5.1	5.2	27.7
C	8	2.4	4.4	7.8	12.0	5.8	8.3	11.5	52.2
	4	2.7	4.0	7.2	11.5	5.4	9.5	11.0	51.3
	6	2.7	4.6	8.3	11.0	6.0	8.8	12.0	53.4

6.6 Mixing patterns

The data given in Table 6.4 were generated by mixing water of set 1 (fresh) with water of set 5 (more saline). The fraction x of type 5 water in each of the samples may be calculated by applying each of the parameters of Table 6.4. For example, applying the K values, the fraction of type 5 water in set 3 is: $1.0x + 0.2(1 - x) = 0.6$, and hence $x = 0.50$; in other words, sample 3 contains 50% of type 5 water and 50% of type 1 water. (Calculate the fraction of type 5 water in set 3, applying the respective concentrations of Ca, Na and Cl. Is the value obtained with K confirmed?) The calculation of mixing percentages is further discussed in the next section.

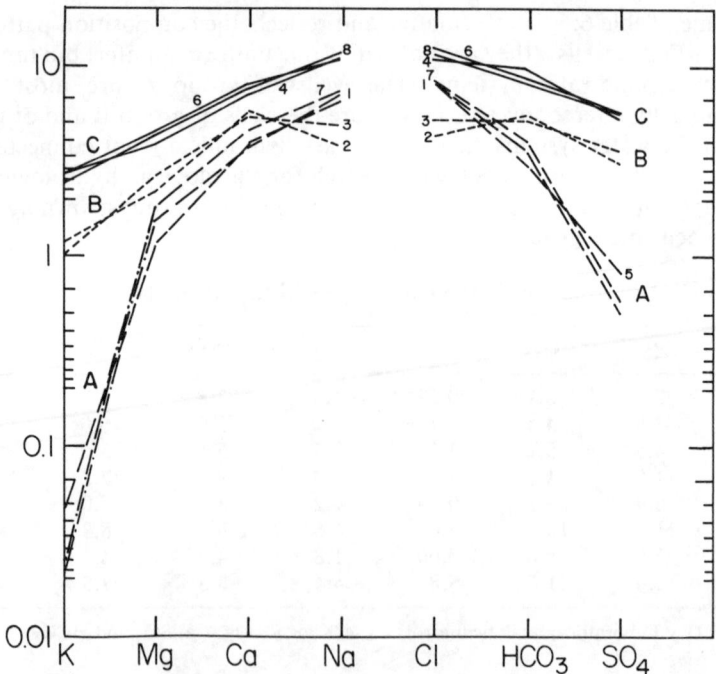

Fig. 6.15 A fingerprint diagram of the data of Table 6.5. Three distinct compositional groups emerge: A, B and C, seen also in Fig. 6.16.

The series of synthetically generated mixtures of two water types given in Table 6.4 has been drawn in Fig. 6.3 as a fingerprint diagram on semilogarithmic paper. The data plot on parallel lines, reflecting the dilution of a saline water type with salt-free fresh water. The same data are plotted on a composition diagram in Fig. 6.17. The data plot on straight mixing lines that extrapolate to the zero points. In contrast, the lines in Fig. 6.6 do not extrapolate to the zero points, indicating mixing of two water types, each containing a significant concentration of dissolved ions. Thus, a glance at a composition diagram reveals:

- Occurrence of mixing (straight lines).
- Dilution of a saline water with fresh water (extrapolation to the zero points).
- Mixing of two water types, both with a significant load of dissolved ions (lines extrapolating to one of the axes).

The patterns in a fingerprint diagram provide the same information:

- Parallel lines (e.g. Fig. 6.3) indicate that dilution occurs.
- Lines with a fan shape (e.g. Fig. 6.5), caused by progressive change in concentrations and relative abundances, indicate mixing of two distinct water types.

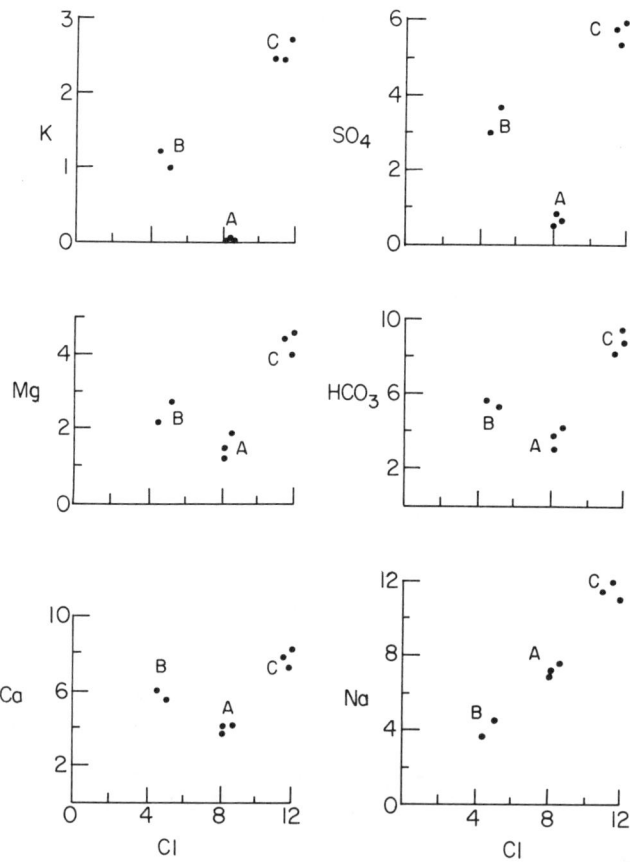

Fig. 6.16 Composition diagrams of the data of Table 6.5. Three distinct compositional water groups emerge: A, B, and C.

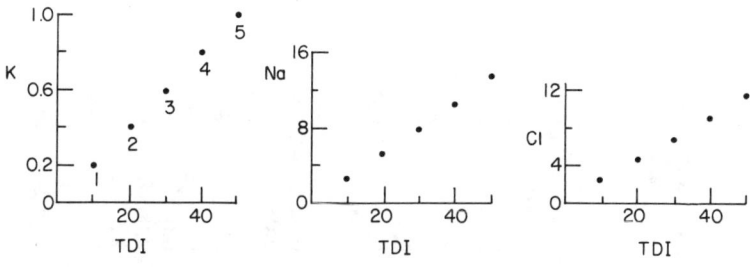

Fig. 6.17 Composition diagrams of the data of Table 6.4 (meq/l). The data have been computed to represent different mixing ratios between water of type 5 and 'distilled' water. The data plot on straight mixing lines that extrapolate to the zero points.

Table 6.7 Dissolved ions in the Kaneh-Samar Springs and Dead Sea brine (meq/l) (Mazor et al., 1973)

No.	Li	Sr	K	Na	Mg	Ca	Br	Cl	SO$_4$	HCO$_3$	Total anions	Total cations	Total ions	Reaction error
1	<0.00014	<0.08	0.2	1.8	3.2	3.2	<0.00001	3.3	0.1	4.92	8.3	8.5	16.8	1.2
2	<0.00014	<0.002	0.2	1.9	3.6	2.8	<0.00001	3.7	0.08	4.49	8.3	8.5	16.8	1.2
3	<0.00014	0.012	0.5	1.9	3.6	3.2	<0.00001	4.3	0.13	4.54	8.9	9.2	18.1	1.6
4	0.001	0.012	0.4	3.3	5.6	4.4	0.02	8.0	0.08	5.23	13.3	13.7	27.0	1.5
5	0.001	0.016	0.8	4.6	6.8	4.0	0.04	10.9	0.55	4.75	16.2	16.2	32.4	0
6	0.01	0.04	1.9	15.2	24.0	8.8	0.32	39.2	1.21	4.82	49.9	49.9	95.4	4.6
7	0.01	0.04	2.0	16.0	22.8	8.8	0.32	39.6	1.02	5.39	46.3	49.6	95.9	3.4
8	0.01	0.04	1.7	16.1	24.8	9.6	0.38	43.4	1.09	4.8	49.7	52.2	101.9	2.5
9	0.02	0.04	2.2	20.0	30.8	12.8	0.52	62.6	1.27	5.28	69.7	65.9	135.6	2.8
10	0.02	0.04	2.7	27.2	34.0	13.6	0.55	64.8	1.47	5.28	72.1	77.6	149.7	3.7
11	0.02	0.04	2.0	19.0	37.2	14.0	0.58	69.7	1.23	5.49	77.0	72.3	149.3	3.1
12	0.03	0.08	3.1	33.9	46.4	14.8	0.79	83.8	1.30	4.52	90.4	98.3	188.7	4.2
13	0.02	0.08	2.4	29.9	45.2	14.8	0.75	87.4	1.82	5.57	95.5	92.4	187.9	1.6
14	0.03	0.08	3.5	32.6	51.6	17.6	0.89	98.3	1.35	5.79	106.3	105.4	211.7	0.4
15	0.04	0.12	3.2	40.7	72.0	22.0	1.29	136.0	1.02	4.66	143.3	138.1	281.4	1.8
16	0.05	0.12	6.4	54.3	94.8	29.6	1.68	181.0	1.73	4.57	189.3	185.3	374.6	1.1
17	0.06	0.16	5.7	65.2	113.2	33.6	2.10	213.0	1.62	4.66	222.1	217.9	440.0	1.0
Dead Sea	2.5	5.9	185.0	1590	3045	687.0	58.0	5486	6.3	3.77	5554.0	5515.0	11069.0	0.4
Dead Sea	2.5	5.9	169.0	1650	3260	756.0	56.0	5814	6.0	3.8	5880	5843	1172	0.3

6.7 End member properties and mixing percentages

Hydrologically deducible end members

A group of springs, Kaneh-Samar, issues at the Dead Sea shores. Results of a hydrochemical study are given in Table 6.7 and plotted in Fig. 6.18, revealing mixing lines. Contributions from the adjacent Dead Sea were suspected. Thus, the Dead Sea values were entered into the Kaneh-Samar composition diagrams of Fig. 6.18. As the Dead Sea values fall on the same lines as the Kaneh-Samar data, the role of the Dead Sea as the saline end member has been established. The procedure has been repeated for four different ions (Fig. 6.18) and the conclusion is, therefore, reached with a high degree of confidence. The fresh end member, in this case, is recharged in the Judean Mountains, with a salinity that is negligible compared with the Dead Sea brine, so one actually talks of dilution.

Examples of hydrologically deducible end members include dams that are potential contributors to adjacent wells, and sea water suspected to intrude into coastal aquifers. Known sources of manmade pollution may be treated as hydrologically suspected end members.

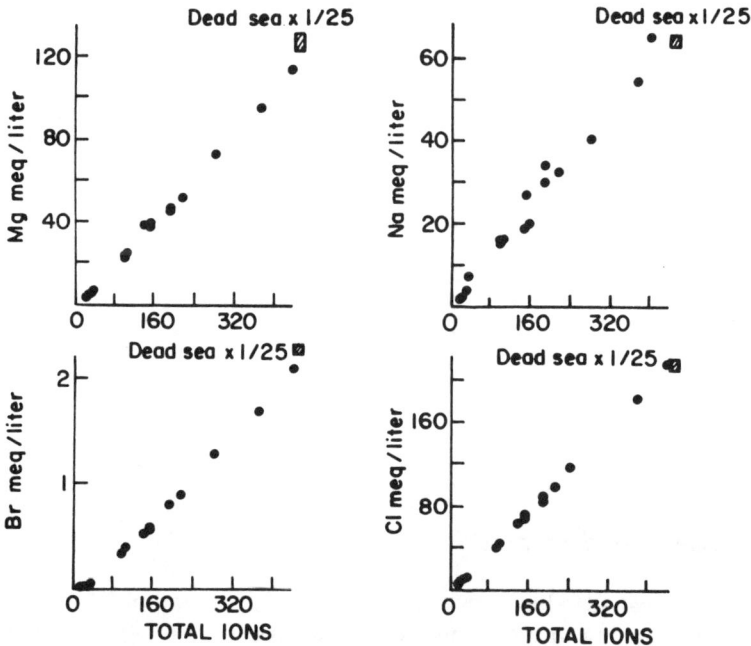

Fig. 6.18 The Kaneh-Samar springs, on the western shore of the Dead Sea. The composition of the Dead Sea water falls on the continuation of the mixing lines obtained by the spring data, supporting the initial hypothesis that Dead Sea water is locally intermixed with fresh water (Mazor et al., 1973). The ion concentrations in the Dead Sea were multiplied by 1/25 to accommodate the Dead Sea in the diagram, which had been designed for the much fresher spring waters.

Calculating mixing percentages

Once the end members are identified, their mixing ratios may be calculated, as discussed in section 6.6. In dealing with real data from complex natural systems, the calculations of mixing percentages warrant some discussion. Table 6.8 contains calculations of the percentages of Dead Sea water in each Kaneh-Samar spring (Table 6.7). It is seen that the percentages derived via the various parameters differ somewhat from each other, due to analytical errors and also possibly, due to superposition of secondary processes. Example: spring no. 15 contains 2.5% Dead Sea water as indicated by the calculation based on the observed concentration of Na (Table 6.8), 2.3% calculated from Mg, 2.4% via Cl and 2.3 via Br, the average value being $2.4 \pm 0.1\%$.

Table 6.8 Percentages of Dead Sea brine diluted by fresh water in the Kaneh-Samar Springs (based on the data of Table 6.7), calculated from the concentrations of various dissolved ions

No.	Na	Mg	Cl	Br	Average
1	0.11	0.10	0.06	—	0.09 ± 0.02
2	0.12	0.11	0.07	—	0.08 ± 0.03
3	0.12	0.11	0.08	—	0.08 ± 0.02
4	0.20	0.17	0.14	—	0.17 ± 0.02
5	0.28	0.22	0.19	—	0.23 ± 0.03
6	0.94	0.76	0.70	—	0.80 ± 0.09
7	0.99	0.72	0.71	—	0.81 ± 0.12
8	1.0	0.79	0.78	0.67	0.81 ± 0.10
9	1.2	0.98	1.1	0.91	1.0 ± 0.10
10	1.7	1.1	1.2	0.97	1.2 ± 0.2
11	1.2	1.2	1.2	1.0	1.2 ± 0.1
12	2.1	1.5	1.5	1.4	1.6 ± 0.2
13	1.8	1.4	1.6	1.3	1.5 ± 0.2
14	2.0	1.6	1.8	1.6	1.8 ± 0.2
15	2.5	2.3	2.4	2.3	2.4 ± 0.1
16	3.4	3.0	3.2	3.2	3.2 ± 0.1
17	4.0	3.6	3.8	3.6	3.8 ± 0.3

Extrapolated end members

Mixing lines can also have negative correlations, for example, for salt-poor warm water mixing with a saline cold end member (Fig. 6.19). In such a case the maximum possible temperature of the warm end member may be deduced by extrapolating the best-fit line to zero TDI. A value of 46°C is obtained in the example shown in Fig. 6.19. The temperature of the true warm end member lies between the warmest measured value and the extrapolated value. In the example given in Fig, 6.19, these two values are

Chemical Parameters – Data Processing

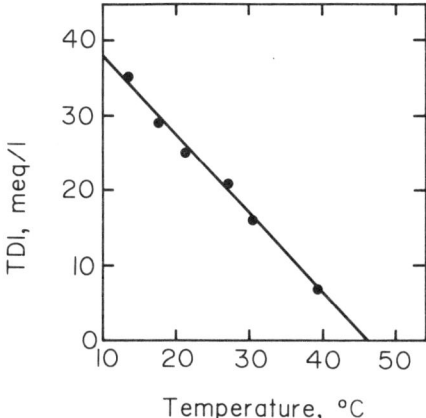

Fig. 6.19 A negative linear correlation between temperature and TDI in a group of springs. By extrapolation, the highest possible temperature value of the warm end member is 46°C. The temperature of the real end member lies between the warmest measured spring and the extrapolated value.

rather close: 39°C and 46°C, respectively. Negative correlations in mixed groundwater systems are often obtained with tritium and ^{14}C data, plotted versus dissolved ions (old saline waters diluted by recent fresh water). The importance of tritium and ^{14}C in this respect will be demonstrated in sections 10.6 and 11.10.

Mixing deduced from 'forbidden' parametric combinations

Dissolved free oxygen is characteristic for well-aerated groundwater, whereas dissolved H_2S is characteristic for anaerobic conditions. Thus, water contains either dissolved free oxygen or H_2S. In special cases both constituents are found in significant concentrations, as a result of mixing.

A second example of a 'forbidden' combination is occasionally observed in groundwaters with little ^{14}C and significant amounts of tritium. The first feature, ^{14}C indicates a great age, several thousands of years, whereas the tritium indicates water recharged after nuclear bomb tests, i.e. post-1953. The use of tritium and ^{14}C as age indicators is discussed in Chapters 10 and 11, and in the present section their use as mixing indicators is considered: water cannot be old and young at the same time, and the combination of little ^{14}C and high tritium can result only by mixing of old and young waters.

6.8 Water–rock interactions and types of rocks passed

The bearing of lithology on water composition has been discussed in section 3.1 and summed up in Table 3.1. Also, the role of soil as a major source of CO_2 has been mentioned in section 1.2, in relation to water–rock interactions induced by CO_2. These points deserve further attention.

Water–rock interactions induced by CO_2

The concentration of CO_2 in the atmosphere is 0.03%. In the soil air, i.e. the air between soil grains, the concentration is 50 to 100 times higher. The source of the extra CO_2 is biogenic: respiration of plant roots and bacterial decomposition of buried plant remains.

Distilled water interacts hardly at all with carbonates or silicates, such as limestone, dolomite and marl, or granite, basalt or sandstone. However, when enriched with CO_2 water turns into carbonic acid and can dissolve rocks, by means of the reaction:

$$CaCO_3 + H_2O + CO_2 \longleftrightarrow Ca(HCO_3)_2$$
limestone   bicarbonate (soluble)

The double-headed arrow indicates that the reaction may go either way: to the right, dissolution of limestone is described; to the left, precipitation of limestone is described. High CO_2 levels drive the reaction to the right (limestone dissolution, e.g. formation of karstic conduits), and depletion of CO_2 drives the reaction to the left (precipitation of calcite and aragonite, e.g. in the formation of stalactites and stalagmites). CO_2 induced limestone dissolution results in groundwaters with Ca as a significant dissolved cation and HCO_3 as a significant anion. The maximum concentration of these ions is limited by the saturation value of calcite in aqueous solutions. Therefore, groundwater in limestone terrains is commonly of good quality.

Carbon dioxide also induces interactions of water with silicate rocks:

$$2KAlSi_3O_8 + 6H_2O + CO_2 \rightarrow Al_2Si_2O_5(OH)_4 + 4SiO(OH)_2 + K_2CO_3$$
K-feldspar   clay   soluble   soluble

$$(Ca, Fe, Mg)(SiO_3) + 2H_2O + 2CO_2 \rightarrow (Ca,Fe,Mg)(HCO_3)_2 + SiO(OH)_2$$
pyroxene   soluble   soluble

Thus, water containing HCO_3 indicates CO_2 induced interactions with rocks, and the balancing cations may indicate the type of rocks passed: Ca may come from interaction with limestone and Ca and Mg together come from interaction with dolomite; K and, even more often, Na in bicarbonate water come from silicate rocks, rich in K or Na feldspars.

Dissolution of evaporites

Water dissolves evaporites, the most common ones being rock salt (NaCl) and gypsum ($CaSO_4$ $2H_2O$). In the first case the groundwater becomes enriched in Na and Cl and in the second case the water becomes enriched in Ca and SO_4.

Evaporites occur as independent rocks, such as rock salt and gypsum, but are common also as veins in marine sedimentary rocks, mainly clay, or as secondary minerals in soils, mainly in arid zones.

The diagnostic value of the dissolved ions and cations

The discussion in the present chapter and in section 3.1 and Table 3.1 reveals the diagnostic value of the ions as indicators for the dominant rock types through which a given groundwater has passed:

Chloride dominated water has passed rock salt (halite) as a rock, or as a guest mineral in other rocks (mainly marine rocks and saline soils).

Sulphate-dominated water has most likely passed gypsum, as a rock or as a guest mineral (mainly in marine rocks and saline soils).

Bicarbonate-dominated water with TDI concentrations of up to about 600mg/l has not passed evaporites, and the nature of rocks interacted with may be deduced from the cations: Ca dominated water has passed limestone; Ca + Mg dominated HCO_3 water is produced by contact with dolomite, Na dominated and K rich HCO_3 waters have interacted with feldspar, plagioclase, and pyroxene, contained in igneous and volcanic rocks.

The discussion of the application of dissolved ions as indicators of the rocks passed by groundwater is of a generalized nature, to show the direction of hydrochemical thinking, useful in establishing constraints, needed to formulate conceptual hydrological models (section 1.4). The topic of chemical water – rock interactions is discussed by Drever (1982), Erikson (1985) and Hem (1985).

6.9 Water composition

Chemical data may be presented in tables and in graphs, but there are also various other ways of describing the chemical composition of water.

Dominant cations and anions. Water may be described by its dominant cation and dominant anion. For example, the water of spring A in Table 6.3 has Ca as the dominant cation and HCO_3 as the dominant anion; the water is of a $Ca-HCO_3$ type. Similarly, the water of spring H in Table 6.3 is of a $Na-Cl$ type.

Order of cation and anion concentrations. A more detailed description of water composition includes the relative abundances of the cations and anions. The example of spring A in Table 6.3 may be described as:

$$Ca > Na > Mg >> K \text{ and } HCO_3 > SO_4 > Cl$$

Chlorinity, total dissolved salts (TDS) and total dissolved ions (TDI). The concentration of the chemical compounds can be expressed in various ways:

- Chlorinity – the concentration of Cl.
- Total dissolved salts or solids (TDS) – the amount of all chemical constituents, determined via electrical conductivity measurements; often called salinity.
- Total dissolved ions (TDI) – the sum of dissolved cations and anions.

Mode of composition description – by equivalents or by weight, per volume water. Each description of water composition must be accompanied by a statement of the units by which the data applied have been expressed, i.e. meq/l or mg/l. This point may be demonstrated by the following set of data:

	K	Na	Mg	Ca	Cl	SO_4	HCO_3	TDI
in meq/l	0.10	5.1	2.5	1.8	5.2	0.3	4.7	19.7
in mg/l	3.9	117	30.2	36.0	185	14.4	225	611

This water may be described in the following two ways:

$Na > Mg > Ca >> K; \quad Cl > HCO_3 > SO_4$ (in equivalents per volume water)

or

$Na > Ca > Mg >> K; \quad HCO_3 > Cl > SO_4$ (in weight per volume water)

It is important to notice that the order of the ions is different in the two modes of expression, and therefore the units applied have to be specified. The same holds true for the chlorinity, TDS, or TDI. The TDI value of the above example is 19.7 meq/l or 611 mg/l.

Water description by its quality or usefulness. Examples:

- *Potable water* – up to 600 mg/l TDI.

- *Slightly saline water*, adequate for drinking and irrigation – up to 1000 mg/l TDI.
- *Medium saline water*, potable only in cases of need, may be used for irrigation of special crops, fish raising is possible, salt concentrations up to 2500 mg/l TDI.
- *Saline water*, adequate for fish raising and industrial use — up to 5000 mg/l.
- *Brackish water*, up to TDI of sea water, i.e. 35 g/l.
- *Brine*, most saline water, with TDI higher than sea water.

6.10 Compositional time variations

The Cl concentration has been repeatedly measured in a well, as shown in Fig. 6.20. What hydrological conclusions may be reached? Two water types, of different Cl concentrations, intermix. What is the nature of this mixing? To answer this question, the nature of the time periodicity has to be discussed. A case study of this kind has been reported by Tremblay et al. (1973) from a coastal well on Prince Edward Island, Canada (Fig. 6.21). The well was pumped daily, from 8 a.m. to 5 p.m., and the Cl concentration was measured at these hours. The chlorinity increased during the day and then dropped until the following morning. This simple series of observations revealed that sea water was drawn into the well due to overpumping, but the inflow of fresh water was sufficient to suppress the sea water intrusion overnight. Hence, if the higher Cl concentration is not wanted, the pumping rate must be reduced. This practical conclusion could be reached by a simple series of repeated Cl measurements, whereas a single analysis could not reveal the dynamics of this water system. It would be desirable to check this conceptual model by repeated measurements of additional parameters.

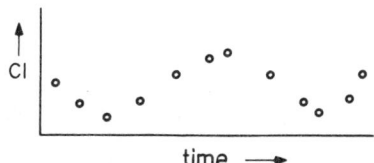

Fig. 6.20 Repeated Cl measurements in the same water source (well or spring). Mixing of two water types is revealed (in text).

Wilmoth (1972) reported Cl measurements in a group of wells in Charleston, Virginia, USA (Fig. 6.22). A gradual Cl increase occurred from 1920 to the end of 1950, when pumping was ended. A check in 1970 revealed that the wells have nearly returned to their original chlorinity. Wilmoth concluded that saline water, underlying the pumped fresh water aquifer, gradually migrated upwards due to the pumping. This is another example of the high value of repeated measurements, further discussed in section 7.6.

This is also a demonstration of the value of historical data, further discussed in section 7.5.

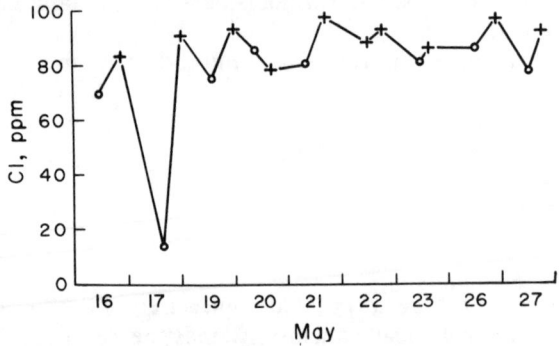

Fig. 6.21 Chloride measurements in a coastal well at Prince Edward Island, Canada. The well was operated daily from 8 a.m. to 5 p.m. and it was analysed at these times: 8 p.m. (o) and 5 p.m. (+). (Following Tremblay et al., 1973). Encroachment of sea water was concluded, with immediate bearing on management (see text).

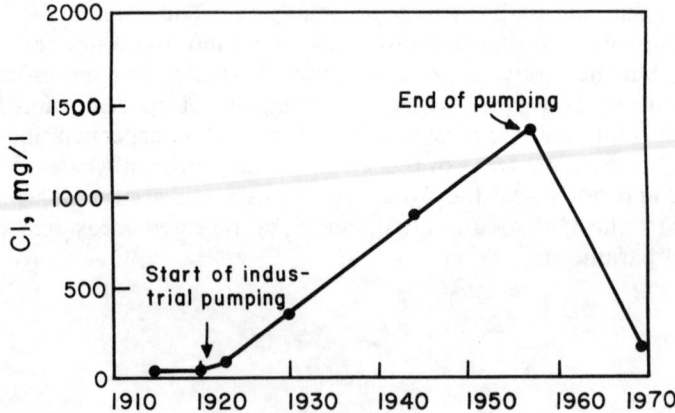

Fig. 6.22 Changes in chloride in groundwater of an overdeveloped aquifer, Charleston, Virginia, USA (from Wilmoth, 1972).

6.11 Some case studies

No case study is a stereotype or a copy of another case. Each case study has its own features, defined by the natural setting and the nature of data obtained. The following case studies were all heavily based on chemical data.

Sea water encroachment

Cotecchia et al. (1974) studied the salinization of wells on the coast of the Ionian Sea in Italy. A fingerprint diagram (Fig. 6.23) served to define a conceptual model. The lowest line (MT) is of a fresh water spring and the uppermost line (I.S.) is of the Ionian Sea water. The lines in between (SR and CH) are of groundwaters with increasing proportions of sea water intrusion. The CH well met the non-diluted sea water at a depth of 170 m. This interpretation seems to be well founded as it is based on six dissolved ions. The whole story is condensed into one fingerprint diagram.

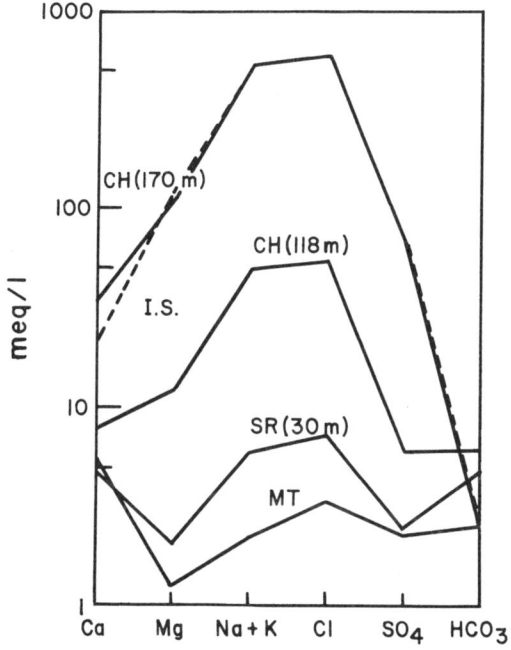

Fig. 6.23 A fingerprint diagram of water in coastal wells, the Ionian Sea (I.S.). MT – a fresh water spring. Well SR has slight contributions of sea water, a feature that is more pronounced in the deeper well CH, which encountered the sea water at a depth of 170 m. (Data from Cotecchia et al., 1972).

Classification into lithologically controlled geochemical water groups

A major application of fingerprint diagrams is in sorting geochemical data into groups. Figure 6.24 represents the results of an extensive study of mineral waters in Switzerland (Vuataz, 1982). Three compositional groups emerged Na – SO_4, Ca(Na) – HCO_3, and Ca-SO_4. A search for

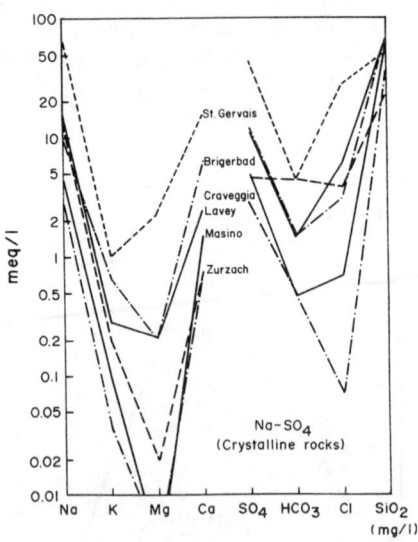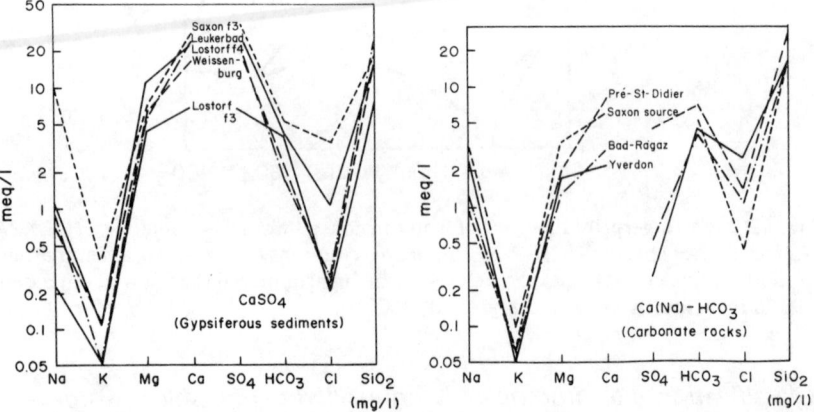

Fig. 6.24 Fingerprint diagrams of data obtained in a study of mineral springs in Switzerland (Vuataz, 1982). Three compositional groups emerged: Na-SO$_4$, Ca(Na)-HCO$_3$ and Ca-SO$_4$. In this case lithology was identified as the major control (see text).

Chemical Parameters – Data Processing

geographical, hydrological, and/or lithological meaning of these compositional groups showed a match with the last factor – lithology: the Na-SO_4 waters issue in crystalline rocks, the Ca (Na)-HCO_3 waters issue in carbonate rocks, and the Ca-SO_4 waters pass gypsiferous sediments. The next step in such a study may be more quantitative, i.e. the conceptual model may be checked and worked out with water-rock equilibration equations and calculation of saturation indices in regards to various mineral compositions. For a detailed discussion of the chemistry of groundwaters the reader is referred to the books by Matthes (1982), Hem (1985) and Eriksson (1985).

Solubility control of groundwater chemistry in an arid region

An extensive study of shallow groundwaters in the Kalahari flatland (Mazor, 1979) revealed a large range of concentrations and different compositions. A composition diagram (Fig. 6.25) produced an evolutionary picture. The concentration of HCO_3 increased with increasing TDI and at 10 meq/l the HCO_3 values levelled off and no further systematic increase in

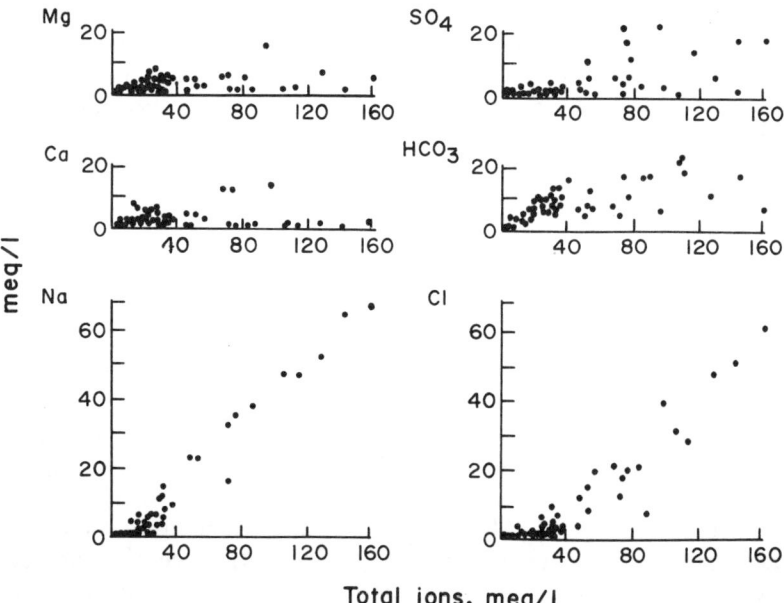

Fig. 6.25 Composition diagrams of an extensive study of shallow groundwaters in the Kalahari flatland (Mazor, 1979). CO_3, SO_4, Ca and Mg increase with increasing TDI in the left part of the diagrams and then level off – at the saturation values for Ca and Mg carbonates and gypsum. In contrast, Na and Cl are low at the lower TDI range and then increase with TDI – indicating that the higher salinization is caused by the more soluble NaCl.

concentration was observed, although the TDI increased significantly. Such a pattern indicates that saturation of a relevant salt controls the concentration of the respective ions. In the Kalahari study, saturation with regard to Ca and Mg carbonates was suggested. This conclusion was supported by:

- Equilibrium calculations revealing that the waters with 10 meq/l HCO_3 were saturated in regards to limestone and dolomite.
- Calcretes ($CaCO_3$ crusts) are common in the investigated area.

The SO_4 concentration is seen in Fig. 6.25 to be low in the range of HCO_3 increase, i.e. up to 40 meq/l TDI. At higher TDI values the SO_4 rises, but levels off at about 20 meq/l, which is the saturation value with regard to gypsum. Na and Cl are low in Fig. 6.25, up to 40 meq/l TDI, and then increase linearly with TDI. This is explained by concentrations of sea borne salts by evapotranspiration. All the waters studied, even the most saline ones, were below saturation in this salt.

The observations discussed in light of the composition diagrams lead to the following conceptual model for the Kalahari groundwaters: infiltrating rain water becomes enriched with soil CO_2 and interacts with feldspars in the covering Kalahari sand. As a result, the water is enriched with HCO_3 that is balanced by Ca, Mg and some Na. In waters that become more saline, saturation of carbonates is reached and calcretes are formed. Another source of salts is rainborne and windborne sea spray, common over all continents. Substantial evaporation causes these salts to concentrate in the soil and they are partially washed down with the fraction of rain water that infiltrates into the saturated groundwater zone. In summary, the evolution of the Kalahari groundwaters is controlled by the solubilities of the relevant minerals, which is:

$$(Ca, Mg)\ CO_3\ <\ Ca\ SO_4\ <<\ NaCl$$

Mixing of cold and warm waters

In a study of the spring complex of Combioula, southern Switzerland, samples were repeatedly measured (Vuataz, 1982) and all the data were summarized in a composition diagram (Fig. 6.26). Mixing lines are indicated by Li, Na, K, Mg, SiO_2, TDS, and temperature, plotted as a function of Cl. Thus, a cold fresh water mixes with ascending warm saline water. Horizontal lines are seen for Ca, Sr and especially for HCO_3, indicating these ions occur in both water end members in the same concentrations. The perfect shape of the Li, Na, K, Mg, HCO_3 and temperature lines in Fig. 6.26 indicates that only one warm water and one cold water are involved in the intermixing. The beauty of this conceptual model lies in its explanation of dozens of different observed water temperatures and ion concentrations as mixing of only two end members.

Chemical Parameters – Data Processing

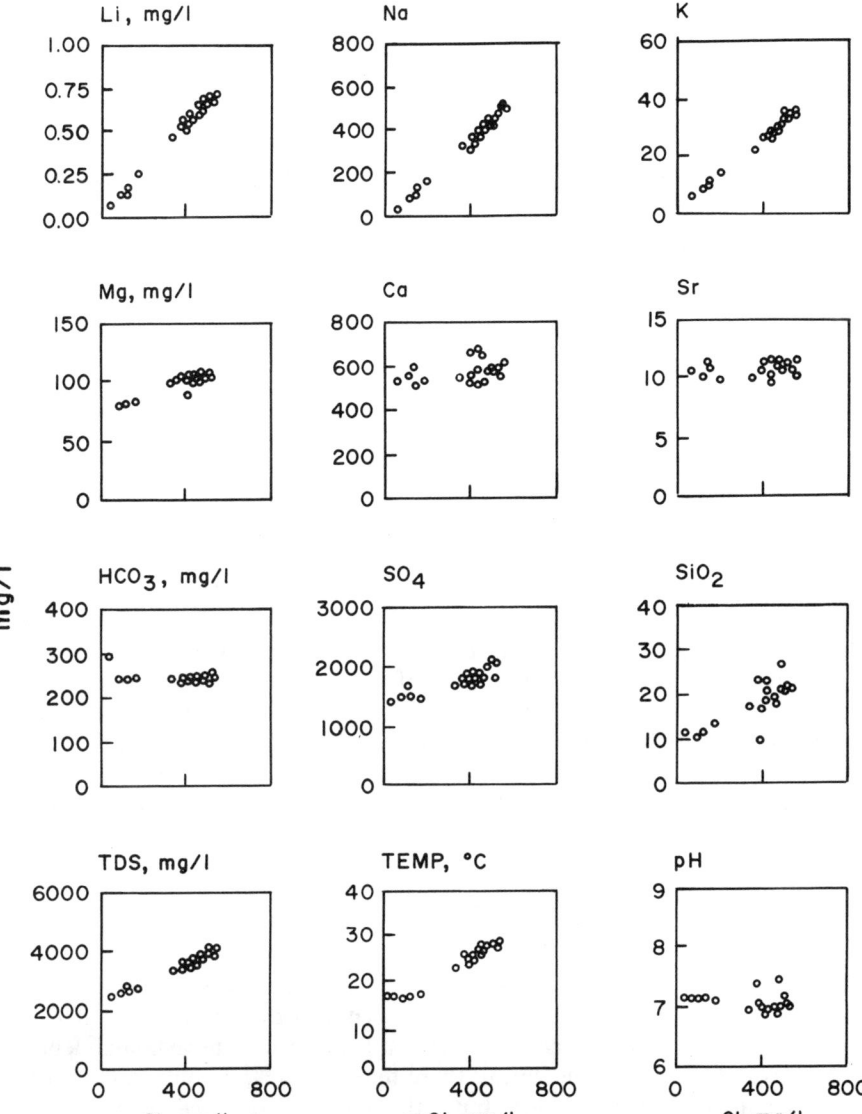

Fig. 6.26 Repeated measurements in a group of springs at Combioula, southern Switzerland (Vuataz, 1982). Positive correlation lines are seen for Li, Na, K, Mg, TDS and temperature, plotted as a function of Cl. Mixing of a cold fresh water with ascending warm saline water is indicated.

7 PLANNING HYDROCHEMICAL STUDIES

A good hydrochemical study is well planned in advance, and as little as possible is left open to luck or fate.

The first stage is a clear definition of the goals or purpose of the study, its extent, and the means available. The answers will differ from one case to another, but there is a common thread: a thorough understanding of the water system under study has to be reached and that has, in turn, to be based on high-quality data, collected in the field or obtained in laboratories. There is always a limit to the number of water sources, wells and springs, that can be included in a study, there is a limit to the number of periodically repeated measurements possible, and there is commonly a strict limit on the type and number of laboratory analyses. Thus, planning of hydrochemical studies is firmly anchored in optimization decisions. The present chapter is devoted to these topics.

7.1 Representative samples

The discussion may be opened with a problem. Temperature measurements conducted at a spring revealed 16°C at the edge of the water body, 18°C at the centre, at a depth of 5 cm, and 21°C at the centre at the maximal depth, which was 1.3 m. Has the spring three temperatures? Is one value 'more representative'? Or should we apply the average measured temperature? The measured temperatures differed substantially, and the cause of these differences has to be understood in order to select the 'right' value. In this example it is clear that the spring water emerged at a certain temperature, but re-equilibration with the ambient air temperature occurred near the surface. This leads to the conclusion that the temperature measured at the maximal depth is closest to the indigenous value. One may generalize: the temperature most different from the ambient air temperature is most representative, and not the average value.

One may go on to ask where in this spring should a sample be collected for laboratory analyses. The temperature may be used as a guide for the location of spring water that has had minimum communication with the surface, i.e. minimum evaporation, oxidation, or incorporation of surface materials. Thus, the most representative sample of the spring is, for all purposes, as deep as possible and as far as possible from the edges. Temperature measurements are a sensitive guide to the most representative spring sample, so they should be conducted first, followed by sample collections and other *in situ* measurements (e.g. pH, dissolved oxygen, conductivity).

Wells also have many 'faces' - a non-pumped well contains a column of water that has possibly interacted with its surroundings and reveals certain temperature values and a certain composition. Upon pumping, water that was shielded in the aquifer enters the well and the properties of the water in the well change. For this reason, water from pumped wells is in many cases regarded to be most representative of its aquifer water. Pumping should be continued until temperature, conductivity, and other parameters measured in the field reach a constant value.

By systematically lowering a temperature sensor into the water standing in a non-pumped well, a temperature profile may be obtained. Commonly the temperature near the water table is closer to the local ambient air temperature; deeper, a constant temperature is obtained which is representative of the aquifer. However, occasionally temperature increases with depth and, in rare cases, temperature reversals may be observed, i.e. a zone of lower water temperature is overlain by a zone of higher water temperature. The latter case indicates lateral flow of the groundwater and differences in rock conductivities. If such a well is pumped, in order to collect samples and conduct field measurements, the values obtained will be a sort of average for the groundwater system being studied.

In practice the various modes of sampling are applied: profiles in non-pumping wells, as well as samples in pumped wells. The mode of sample collection has to be taken into consideration at the data processing stage.

Over-pumping of a well may change the local pressure distribution in the water system, to the extent that water from an adjacent aquifer may breach in and change the water properties. So, over-intensive pumping may introduce new complications. An obvious example is encroachment of sea water into coastal wells.

The question of representation of water samples collected at wells has also to be addressed in light of the casing perforations. Well casings may be perforated along the entire section of the saturated zone, or they may be perforated only at certain intervals. Thus, a knowledge of the perforation geometry of a well is needed to understand what a collected sample represents.

The discussion so far may create the impression that representative water samples are out of reach and nothing can be done about it. In fact, simple precautions provide reasonable solutions:

- The mode of data and sample collection in the field has to be written down and taken into account at the data interpretation stage.
- Samples protected from surface interaction are preferable.
- In non-pumped wells, measuring of profiles is recommended.
- In pumped wells, samples should be collected only after stabilization of properties has been achieved. The relevant pumping history has to be documented.
- Periodically repeated data collection is desirable.

Confirmation of previous data indicates good representation of all data sets. Variations in the periodical measurements call for careful sorting of the most representative values (which necessitates understanding of the causes of the observed variations).

7.2 Data collection during drilling

The fingerprint diagram shown in Fig. 7.1 depicts gradual salinization with depth at the Amiaz 1 well, west of the Dead Sea (Mazor et al., 1969). Fresh water was encountered at a depth of 32 m, whereas saline water, of the local Tverya-Noit group, was found at a depth of 85 m. A practical consequence of this is that some fresh water may be abstracted from a depth of 30-40 m. The example of the Amiaz 1 well may be generalized: in each area several water bodies may be passed by a drill. All of them should be documented in the driller's records, all should be measured *in situ*, and samples should be collected for laboratory measurements.

The study of water encountered during drilling causes technical difficulties: most important are interferences of water introduced by the drilling procedure, difficulties in noticing natural water horizons that are passed by the drill, and the cost of stopping the drilling operation for measurements. These difficulties may be overcome in the following ways:

- Use of every interruption in drilling (e.g. weekends or mechanical breakdowns) to measure the water table, temperature, and conductivity. Repetition of these measurements may reveal progressive changes, indicating restoration of the water system. At the end of each break, water samples should be collected for laboratory analyses. A sample of the water applied for the drilling should be analysed as well. Differences between the latter and the well samples will indicate the water bodies encountered and their properties.
- Water in the drill hole should be removed with a bailer, i.e. a water sampling rod, lowered and lifted with the drilling equipment. The amount of water to be removed should be several times the volume of the water-filled section of the drill hole. The bailer also serves to collect the samples. If the water is quickly transferred into bottles, the temperature may even be measured in them. However, lowering a temperature logger, a pump and a special sampling device into the drill hole is preferable.

Planning Hydrochemical studies   105

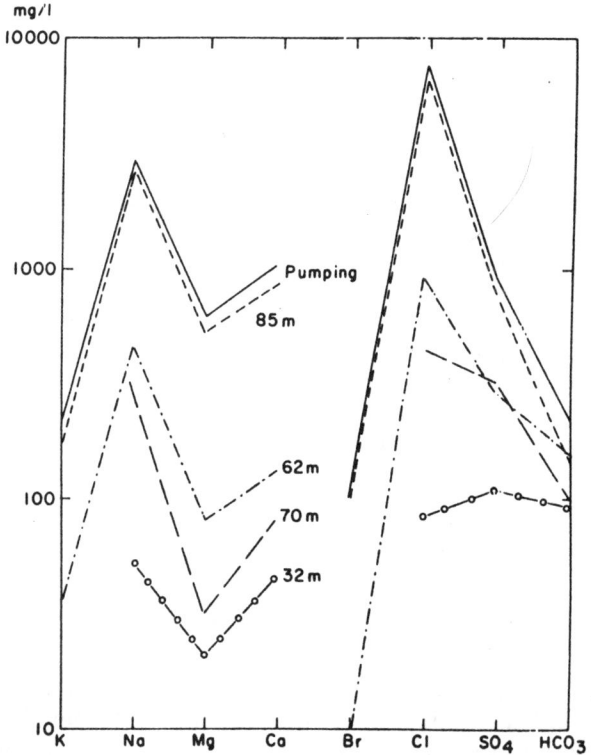

Fig. 7.1 A fingerprint diagram of water from various depths, collected during drilling of the Amiaz 1 well, west of the Dead Sea (Mazor *et al.*, 1969). Several water horizons with different water qualities were encountered.

- Whenever possible, dry drilling (with compressed air) is preferable.
- The budget for drilling a well should include the costs of several stops, required to measure water bodies that may be passed.

Information obtained during drilling is most valuable. Certain drill holes may never be used as wells and the only outcome is the information gained during drilling. Yet, in most cases this information is not collected.

Measurements during drilling provide the hydrological information in the depth axis. Therefore it is essential that a hydrochemist performs detailed work during each drilling operation. Thus, the planning of hydrochemical studies should routinely include drilling operations.

7.3 Depth profiles

Figure 4.21 depicts temperature profiles in a well close to the Mohawk River, New York, USA. In May a temperature of 43°F has been observed at the measured depth interval of 205–165 feet above sea level. The July profile looked totally different: it showed 57°F at the top, 64°F at the centre, and 52°F at the bottom. The January profile looks different again: 56°F at the top, 68°F at the centre, and 52°F at the bottom. The interpretation of these temperature profiles is discussed in section 4.8. For our present topic, the planning of hydrochemical studies, it is important to keep in mind that water strata are not necessarily uniform – if they vary in their temperature profiles, they may vary in other parameters as well.

A sample collected from a pumping well provides average water properties and the detailed profile provides the fine structure, which is most important to the understanding of the dynamics of groundwater movement, as discussed in section 4.8.

Depth profiles in observation wells, or non-pumped wells, are conducted by lowering sensors into the well. The depth is usually read from markings on the electrical cable. The lowering of sensors into the well agitates the water and mixes the water column in the well. For this reason:

- Measurements are taken downwards, so the sensor enters non-disturbed parts of the water column in the well.
- Several parameters are measured together e.g. temperature, conductivity, and pH.
- Water level measurements, which also agitate the water column, should be done last.

Samples may be collected from a depth profile by carefully lowering a cylindrical sampling bottle (commerically available). It should be gradually lowered to deeper sections so that each bottle will collect non-disturbed water. The depths of perforated sections and other details of the well construction have to be known for proper interpretation of the data obtained in each profile.

Planning of a hydrochemical study has to take into account the need to measure depth profiles and to select the proper time to do them.

7.4 Data collection during pumping tests

Section 4.6 is devoted to chemical and physical measurements during pumping tests. The conclusion reached is that physical and chemical follow-up of pumping tests is essential for their proper interpretation. Pumping of one water body (aquifer) is commonly *assumed*, but physical and chemical measurements can *establish* the number of water bodies affected by the pumping. Thus, the hydrochemist should participate in pumping tests and in the interpretation of their data. The equipment needed includes probes for measuring conductivity and temperature, to be placed at the pump outlet to ensure minimum temperature re-equilibration. Samples for dissolved ions, stable hydrogen and oxygen isotopes, and tritium measurements should be taken before pumping is commenced, or immediately at the beginning, and then at intervals up to the end of the pumping. About 10 samples should be taken: the intervals may be planned in light of the pumping test program. The first and last samples should be sent first to the laboratories. If they reveal the same concentrations, and provided the temperature and conductivity remained constant during pumping, then pumping of one water system is demonstrated. But, if conductivity and/or temperature changed during pumping, or the first and last sample were found to differ in their properties, then the rest of the samples have to be sent for analyses. Modes of data interpretation are discussed in section 4.6.

7.5 Importance of historical data

The data collected during the drilling of the Amiaz 1 well (section 7.2 and Fig. 7.1) were put aside once the drill entered saline water and was abandoned. Several years later a water source of low output but good quality was needed in the area, and on that occasion all historical data were scanned and studied. The existence of the required water body was spotted from the historical data, at no extra cost. A study of groundwater systems along a transect through the Judean Mountains, central Israel, was recently performed (Kroitoru et al., 1987; Kroitoru, 1987). Besides many new measurements, all available data from published reports and non-published archives were incorporated. One measurement turned out to be of special value: post-bomb tritium was observed in the Elisha Spring, situated on the eastern end of the Judean Desert, 22 km east of the recharge area, which is around Jerusalem. The point of importance was that the tritium was observed as early as 1968, at most 15 years after the onset of nuclear bomb tests, held between 1953 and 1963. Thus, it turned out that groundwater flow to the Elisha spring was at least $22:15 = 1.5$ km/y which, in turn, indicated flow in karstic (old) conduits. The historical data provided much additional information in this study and turned out to be most useful (Kroitoru, 1987).

Many researchers tend to disregard data obtained by previous workers and prefer to concentrate on their own new measurements. Explanations given are that the old data are incomplete, the laboratories were not that good in the old days, or the old data are not available. These arguments may easily be countered:

- Many of the previous researchers produced excellent data.
- The quality of the old data can be assessed by processing techniques discussed in section 5.10.
- Archive data are available from local water authorities.
- Historical data have the advantage of being available free of charge.
- The old data are available in the early stages of a new study and can help planning the new research campaign.

It is assumed that the reader is now convinced that historical data are useful. But the most important argument has not yet been spelled out: the old data were collected when the water systems were less disturbed by human activity. Thus, historical data record information that cannot now be collected, and serve as a database, needed to understand the evolution and dynamics of studied water systems.

Historical data enlarge the time span of repeated observations, the topic of the next section.

7.6 Repeated observations or time-data series

Repetition of field measurements and laboratory analyses is an essential part of hydrochemical studies. Sources of recharge vary over the year, e.g. summer rains, winter rains and snow melt. These are distinguishable by their isotopic (and occasionally also chemical) composition (Chapter 9), and their seasonal inputs may be established provided the relevant water systems are monitored over at least one year. Water table and discharge responses to precipitation, or flood events, provide indications on hydraulic interconnections. Such responses can be detected only by comparing the results of repeated observations. Chemical variations and temperature fluctuations are indicative of mixing processes, thus justifying the effort invested in repeated observations. Last, but not least, the man-induced changes in water systems can be understood by comparing data from repeated measurements. These provide indications of deterioration caused by harmful human impacts, or indications of improvements achieved by proper management.

Having praised repeated measurements, we have now to discuss their incorporation into the planning of hydrochemical studies. A common rule of thumb is that each water source should be studied four times, to cover at least one hydrological cycle, including the four major seasons. This may be regarded as a minimum, because time-data series with only four points suffer from poor resolution.

More data points, i.e. more repetitions, were needed in the hydrographs of wells depicted in Figs 4.7–4.10, the temperature observations plotted in Figs 4.19–4.23, or the discharge, temperature and Cl concentration values discussed in the time series of Figs. 4.24–4.27. Thus, in many studies monthly measurements are needed for at least 14 consecutive months in order to fully cover a hydrological year. Repeated studies are especially necessary in areas that are scarce in accessible wells or springs. When working on single water sources, multisampling provides an insight into the underground plumbing, establishing existence of a single water aquifer, several waters that intermix (sections 6.4, 6.5, and 6.6) and conduit-dominated versus porous medium recharge intake (sections 2.2 and 2.9).

Thus, measurements may have to be repeated once in every few years (Fig. 6.22), seasonally, monthly, daily (Fig. 6.21) or hourly (pumping tests).

Planning Hydrochemical studies 109

In the absence of repeated measurements the dynamics of a system may be revealed by measurements performed in many sources (wells or springs) in the studied area (Figs. 6.2 and 6.25).

7.7 Search for meaningful parameters

Let us have a look at the compositional diagrams of the Combioula study (Fig. 6.26). Suppose a new study is planned, to establish the impact of new wells on the warm mineral springs, and suppose that for budgetary reasons the new study has to be based on repeated measurements of only three parameters. The detailed study portrayed in Fig. 6.26 revealed that the dominant process in the Combioula spring complex is intermixing of warm saline water with cold, relatively fresh, water. The three parameters to be included in the new study should be reliable tracers of the mixing. Thus, they have to be selected from the parameters that provide good mixing lines in Fig. 6.26, namely Li, Na, K, Cl and temperature. Parameters that have equal values in both end members are useless (these include in our example Ca, Sr, HCO_3 and SO_4). Parameters revealing large data spreads are also useless, e.g. the SiO_2. By this method of sorting of parameter efficiencies we are left in the Combioula case with five parameters: Li, Na, K, Cl, and temperature. Temperature, being a physical parameter (with short memory, as it readily re-equilibrates) should be included among the most meaningful parameters. The remaining parameters should include at least one anion, Cl, and one cation, perhaps Na which is dominant and easy to measure.

The selection of meaningful parameters is essential for economical reasons: focusing of efforts and means on the application of a small number of selected, most promising, parameters.

The search for meaningful parameters necessitates that each study has a preliminary stage at which a large number of parameters is measured. From the preliminary results the most informative parameters may be selected. This first stage of measurement of a large number of parameters is essential, as no case study resembles previous ones, and the list of meaningful parameters varies from case to case: temperature measurements are essential in warm water systems and less informative in cold systems with constant temperatures, Cl is less important in many fresh water systems but essential to trace brines; in contrast, HCO_3 is a sensitive tracer in fresh waters. Stable isotope measurements are meaningful in aspects that are different from the meaning of dissolved ions (e.g. altitude of recharge, identifying evaporation brines, or indicating a thermal history). Hence, if money is tight, the list of dissolved ions measured may be cut in order to save the stable isotopes. Tritium is an essential age indicator for post-1953 water, but it may be dropped if preliminary work reveals that the study deals with older waters. Carbon-14 measurements necessitate handling of hundreds of litres of each sample, and their determination is costly. Therefore, although they are a must in the preliminary study stage, they may be reduced once the basic conceptual model is reached.

The last sentence provides a guideline to the timing at which the number of parameters measured in a study can be cut: when an initial conceptual model is reached. Fewer parameters have to be measured to check the validity of the model, or to follow the evolution of a system due to exploitation.

7.8 Sampling for contour maps

Extensive hydrological studies, extending over large areas and with large numbers of accessible wells, pose severe difficulties of presenting and processing the wealth of data obtained. One mode is to produce maps with contour lines of equal values of the various parameters.

Planning of a hydrochemical study should include a clear decision – will the data be expressed in contour maps? If so, data have to be obtained with an adequate geographical coverage and adequate density of measured points. These requirements are necessary to ensure adequate resolution and to avoid 'white holes' in the maps.

Hydrochemical contour maps should be prepared from data collected at the same time. Mixing of data from different seasons, or even different years, reduces the meaning of the map. On the other hand, comparison of contour maps of data obtained at different dates may be most informative. The need for many sampling points may result in a need to cut other efforts, e.g. to reduce the number of measured parameters, as discussed in the previous section.

7.9 Sampling along transects

Sampling at wells and springs lying along a selected transect facilitates data processing and presentation. An example from a recent study in the Judean Mountains is given in Fig. 11.17.

A hydrochemical transect has the advantage of graphically linking parametric variations with geographical locations (topographic transect), geological features (geological transect), and hydrological setting (water table transect), as seen in Figs. 4.5 and 4.6.

Planning a hydrochemical study should include the question of whether the transect approach is desired. If so, the locations of the transect should be decided. Such a decision should take into account:

- The geological structures (a transect crossing major structures is desired).
- Availability of accessible wells and springs.
- Priority to wells and springs of large discharge (best representing the studied water system).
- Even distribution of sampling points along the transect.

Planning Hydrochemical studies 111

7.10 Reconnaissance studies

The term reconnaissance study relates to studies conducted for the first time, covering extended areas, to be completed fast and with minimal costs. The purpose of reconnaissance studies is to get a general picture of the water systems involved, to locate promising sites at which large amounts of high-quality water can be found, to scan the area for features of special interest, to locate possible pollution processes, and to arrive at a conceptual model. The results of reconnaissnce studies are the basis for the planning of detailed studies in selected areas. The major points characterizing reconnaissance studies are the inclusion of maximum available number of wells and springs, and the measurement of as many parameters as possible. The first point is needed for representative samples, and the second point is necessary in order to find out those parameters that are most informative in the studied system.

In planning a hydrochemical study, a clear distinction has to be made between a reconnaissance study and a detailed study.

7.11 Detailed studies

A detailed study is characterized by:

- Relatively small study areas.
- Existence of former data – historical or of a preliminary study.
- Requirements for thorough research that should provide answers to specific questions.
- A quantitative approach, whenever possible.
- Inclusion of observations needed to check a proposed conceptual model.

Detailed studies are conducted over a period of a year or more, and include adequate repetitions of field measurements and sample collections for laboratory analyses. These studies are based on a variety of sampling strategies, e.g. repeated sampling in springs and wells, depth profiles in non-pumped wells, search for water occurences during drilling, and measurements during pumping tests.

The number of parameters measured will be large at the beginning of a detailed study but it will gradually narrow down as the system is better understood and the most informative parameters are spotted.

7.12 Summary: a planning list

It is said that a well formulated question contains half the answer. By the same token, a well planned study is a key to success of the study. The following list of planning stages is suggestive and is given as a summary of the present chapter:

1. Definition of the purpose of the study – practical and scientific.
2. Definition of the study area.
3. Collection of all available data (geographical, hydrological, geological, temperatures, chemical and isotopic compositions, time variations and so on).
4. A conceptual model, or several possible models, suggested by previous and new workers, developed on grounds of the available data.
5. Definition of the nature of the study – reconnaissance or detailed.
6. Sampling strategy – sampling for a contour map or sampling along a transect.
7. Preparation of a list of wells, springs and surface water bodies to be included in the study (in light of points 1–6 above). Total number of sampling points should be estimated.
8. Listing locations for special measurements and sample collections (e.g. depth profiles, sampling during drilling, and during pumping tests).
9. Listing parameters to be measured in the field and type of instruments needed.
10. Listing parameters to be analysed in laboratories, including names of the laboratories and types of sampling vessels needed.
11. Frequency of planned repetitions of field measurements and laboratory analyses.
12. Potential bottlenecks that need special care, e.g. access to wells on private ground, long waiting time in a special laboratory, or co-operation of drilling authorities.
13. Schedule and timetable.
14. Budget requirements, based on the above points and ordered by nature of expenditure: salaries, means of transportation, equipment, sample collection vessels, laboratory fees, maps and photos, miscellaneous items.

8 CHEMICAL PARAMETERS – FIELD WORK

8.1 Field measurements

Certain chemical and related physical parameters have to be measured at the well or spring site. The list of field measurements includes parameters that change in stored samples and cannot be measured in the laboratory in a meaningful way. Most conspicuous in this respect is temperature, but pH, alkalinity, and dissolved O_2 may also change between sampling and analyses in the laboratory.

For each parameter there exists in the market high quality instruments, accompanied by instruction manuals, describing the operation of the sensors and calibration and measurement procedures. Portable versions are available, with a short cable for the measurement of water of springs or water pumped from wells. Instruments for well logging (depth profiles) are available with long cables in portable versions, suitable for surveys, and in fixed modes appropriate for continuous monitoring.

Instrument sets are available that measure several parameters simultaneously, a procedure that is time-saving and convenient.

A comprehensive and easy to follow text on sampling procedures and biological, chemical, and physical measurements in the field has been prepared by Hutton (1983).

Field measurement instruments should be checked for their limits of detection, accuracy, and reproducibility, and their measurement errors should be established. This may be done with standards, duplicate samples, and dilutions of samples, similar to the checks discussed for laboratory measurements (section 5.9). Such checks should be performed as part of the preparations for the field work, and the results should be recorded and reported with the data (sections 5.7 and 5.8). Checking the field equipment in advance is essential to secure proper operation in the field.

Field measurements are needed for a number of purposes:

- Measurement of parameters that change, or may change, after removal of the water from the sampling point.
- Immediate provision of information on water quality, needed to decide extent of sample collection.
- Providing checks on the laboratory data. For example, disagreement between electrical conductance measured in the field, and TDI determined in the laboratory, indicates erroneous field measurements, erroneous laboratory results, or mislabelling of samples. Agreement between field and laboratory data raises the confidence in the data and indicates that no secondary processes have occurred between sampling and laboratory measurement.

8.2 Smell and taste

On approaching a water source one occasionally senses an unpleasant smell. In most cases this is caused by sulphur compounds, mainly H_2S. This information should be recorded in the field notes, and taken into account at the stage of data interpretation, because the presence of H_2S and related compounds indicates:

- Reducing conditions.
- Most probably bacterial activity.
- Possible occurrence of sewage or other pollutants.

Human ability to smell H_2S is limited to the ppm range. At higher concentrations the odour is not noticed. Hence, if a H_2S smell is noticed while approaching the site but disappears at the well or spring itself, a high H_2S concentration should be suspected and relevant sampling is recommended.

Smell is a crucial factor in the ranking of water quality. In some cases aeration of water supplies is practised in order to get rid of compounds causing unpleasant smells.

Tasting water provides immediate quality information, for example:

- Estimate of the total salinity.
- Identification of major ions, e.g. NaCl tastes 'salty', $MgSO_4$ is bitter, and iron 'catches' the mouth in an unpleasant way.
- Identification of pollutants, e.g. sewage or industrial waste.

Tasting ability may be developed by tasting waters of known compositions. The information gained by tasting is of immediate use in the field: different water types can be recognized, and more detailed sample collection may be decided for special cases.

A word of warning is needed at this point: do not taste water if you suspect it to be contaminated.

8.3 Temperature

Potential uses of temperature data have been discussed in sections 4.7-4.9. The ease with which temperature is measured, and the benefits to the understanding of groundwater systems, make the measurement of this parameter a 'must' in every study. This is also true for repeated measurements, depth profiles, and pumping tests.

Temperature may be measured by mercury-filled thermometers or by thermistors read digitally. Resolution and accuracy of $\pm 0.1°C$ are desirable. Thermometers should be calibrated by comparison with standard thermometers, or by immersing in fresh water – ice mixtures ($0°C$) and in fresh boiling water (the local boiling point is altitude-dependent).

Temperature data are needed for water-rock equilibrium calculations, as well as for the indentification of water groups, the determination of water end member properties (section 6.7) and to deduce depth of water circulation (section 4.7).

8.4 Electrical conductance

An electrical current can be carried through water by the dissolved ions. Electrical conductance is expressed in units of micromho/cm, also called microsiemens (μS). A positive correlation is observed between the electrical conductance and the total dissolved ions, TDI. This correlation is linear up to 500 meq/l of TDI. At higher concentrations of dissolved ions the line in a conductance – TDI diagram levels off.

The electrical conductance of a given water solution increases with temperature. Field probes of electrical conductance are therefore temperature-compensated. Conductivity values obtained in the field should be plotted against the corresponding TDI concentrations measured in the laboratory. A good linear correlation confirms the high quality of the data. Outstanding values should be suspected as erroneous, and should be discussed with the laboratory staff for possible detection of errors or repetition of measurements. Conductivity measurements are of special use in the following cases:

- Monitoring pumping tests for the detection of possible intrusion of different types of water (section 4.6).
- Measuring depth profiles in non-pumped wells, to search for different water strata (section 7.3).
- Monitoring water flowing from a well, opened just before sample collection: samples may be collected once a constant conductivity value is obtained.
- Continuous monitoring in observation wells to detect arrival of rain and flood recharge.

- To substitute for TDI concentrations that are needed for data processing, but are occasionally missing in cases of incomplete laboratory analyses. The needed TDI value may be read from the conductivity - TDI curve of the studied system.

8.5 pH

The pH of a solution indicates the effective concentration of the hydrogen ion, H^+. The units of pH are the negative logarithm of hydrogen ion concentration, expressed in moles per litre:

$$pH = - \log H^+$$

The pH describes the composition of water: pH 7 indicates neutral water, lower values indicate acid water, and higher values indicate basic water.

The pH is controlled by various reactions and the presence of different compounds. In fresh water systems the carbonate system $CO_2 - HCO_3 - CO_3$ plays a primary role in determining the pH. In other cases presence of H_2S, or its oxidized form, sulphuric acid, determines low pH.

A knowledge of the pH, up to ± 0.1 unit, is essential for water-rock equilibrium calculations. For such calculations experts may be consulted, but accurate pH data have to be supplied by the field hydrochemist.

The pH measured in the field should be compared to the pH measured in the laboratory. Agreement between the two values indicates that no secondary reactions have occurred in the sample bottle, e.g. loss of CO_2 or biological activity. Whenever the two pH measurements agree, the water chemistry measured in the laboratory is relevant to the water *in situ*. However, deviations of the laboratory pH from the pH measured in the field indicate either that one of the measurements is wrong, or that secondary reactions have occurred in the time elapsed since the samples were collected. In such cases the preservation of the samples has to be checked and improved (section 8.10), and the period between sampling and laboratory measurements has to be shortened.

Disagreement between field and laboratory pH measurements places a question mark over the reported concentration of biologically involved ions such as HCO_3, NH_4, SO_4, H_2 and Ca. On the other hand, the reported concentrations of more conservative ions may be correct, e.g. Cl, Br, Na, K or Li.

The pH value measured in the field is an important parameter in water quality assessment in relation to corrosion problems and taste.

The pH is determined by means of a glass hydrogen ion electrode of known potential. Calibration is carried out with special solutions of known pH.

8.6 Dissolved oxygen

Rain and surface waters equilibrate with air, becoming saturated with dissolved oxygen. Equilibration with soil air goes on until water reaches the saturated zone, at which it is isolated from further contact with air.

The concentration of dissolved oxygen in air-saturated water depends on:

- Pressure, which is controlled by the altitude.
- Temperature, increasing temperature decreasing the oxygen concentration.
- Salinity, causing a decrease in the dissolved oxygen (negligible for fresh waters).

The initial concentration of dissolved oxygen in recharged water can be computed for each study area. The concentration at sea level (1-atmosphere), as a function of the average annual local temperature, can be read from Fig. 8.1. At locations with altitudes above or below sea level a correction factor has to be applied to the oxygen concentration read from Fig. 8.1. The altitude correction factors are given in Fig. 12.2. For example, air-saturated water at sea level at 17°C contains 9.0 mg/l oxygen, and at an altitude of 330 m it contains $9.0 \times 0.96 = 8.6$ mg/l of O_2. Comparison of the calculated initial O_2 concentration with the value measured in the field reveals the fraction retained. The missing O_2 has been consumed by oxidation of rocks and biological activity. The rate of oxygen consumption

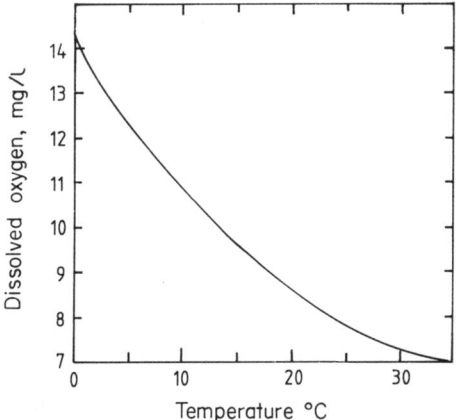

Fig. 8.1 Dissolved oxygen (mg/l) in fresh water, as a function of the ambient temperature.

depends on the aquifer lithology, e.g. occurrence of pyrite and other oxygen-consuming minerals, and availability of organic compounds and nutrients needed for O_2 consuming biological activity. The consumption of dissolved O_2 serves also as a semi-quantitative age indicator: older water tends to have lost most or all its of dissolved oxygen. Generally speaking, in phreatic aquifers water retains a significant portion of the initial O_2, whereas in confined aquifers little or no O_2 is retained.

The determination of dissolved O_2 is informative in several respects:

- Defining aquifer environment in terms of degree of anaerobic conditions.
- Detecting changes occurring in a water system: decrease in dissolved O_2 may indicate increase in biological activity, connected to arrival of sewage or other polluting fluids.
- Identifying systems with dominant conduit-controlled recharge and underground flow. These have limited water-rock contact and fast flow, and retain their initial dissolved O_2 even at large distances from the intake zone.
- Identifying confined systems of old water, likely to have lost their O_2.
- Identifying intrusion of water from adjacent wells, as a result of pumping tests, and in intensively exploited well fields.
- Identifying different water strata in depth profiles.
- Identifying mixing of different water types by the occurrence of 'forbidden' ion combinations, e.g. dissolved O_2 and H_2S.

The concentration of dissolved oxygen is determined by a set of two electrodes immersed in an electrolyte, contained in a vessel with an oxygen-permeable membrane wall. The cell is placed in the container through which the measured water flows. Oxygen diffuses into the cell, in direct proportion to its concentration in the measured water. The diffused oxygen is consumed by the cathode, creating a measurable current which is proportional to the diffused oxygen and, in turn, to the O_2 concentration in the measured water.

The instruments measuring dissolved O_2 are calibrated by immersion in aerated water samples, for which the dissolved O_2 is computed from given curves that take into account the water temperature and salinity, as well as altitude.

8.7 Alkalinity

Alkalinity is an expression related to the sum of the carbon containing species dissolved in the water: CO_2, HCO_3 and CO_3. Bicarbonate is commonly the dominant ion.

Alkalinity has to be measured in the field because CO_2 is often pressurized in groundwater, compared to air-saturated water at the surface (due to addition of CO_2 from soil and other underground sources). Upon exposure to the atmosphere some CO_2 may leave the water, causing part of the HCO_3 to break down. For these reasons it is highly recommended that alkalinity is determined in the field. The field-measured alkalinity values are needed for water-rock saturation calculations. Various set-ups are available for alkalinity measurements in the field, by titration of the sample with an acid and pH colouring indicator.

8.8 Sampling for dissolved ion analyses

Sampling procedures for the determination of dissolved ions should be discussed with the relevant laboratory. Commonly, half-litre plastic bottles are adequate. Occasionally a separate bottle is recommended to which a few drops of pure acid are added to secure low pH, in order to prevent precipitation of some ions, e.g. as carbonates. For SiO_2 analyses, addition of known amounts of distilled water to the sample is recommended to avoid precipitation of silica as a result of cooling of water.

Separate samples are commonly collected for H_2S determination. To these a fixating agent is added, each laboratory preferring its own material.

The details of a sampling expedition have to be worked out with the laboratory staff in advance, and the proper (clean) bottles have to be obtained. In addition to being clean, bottles have to be rinsed in the sampled water.

Samples have to be well labelled in the field – never rely on your memory in this respect.

Filtering. Water samples contain dispersed solid particles. These may be removed by filtration, either in the field or in the laboratory. Some workers carefully filter every sample in the field in adequate quantities, but others argue that the water emerges with particles that are part of the hydrological system under study and therefore their filtration can be postponed until arrival at the laboratory, or skipped.

Filtering needs experience; non-experienced personnel may introduce contamination and cause fractionation. The topic should be discussed with the laboratories concerned.

8.9 Sampling for isotopic measurements

The nature of isotopic data is discussed in the following chapters, but the sample collection techniques belong to the present chapter.

For δD and $\delta\ ^{18}O$ measurements, 50 ml samples are adequate. Evaporation of samples has to be avoided by prompt closing of the bottles and storage in the shade. Normally each laboratory prefers its own bottles.

For tritium, 1 l bottles are desirable, to allow for repeated measurements and pre-concentration (in the case of low tritium concentrations). Glass bottles are preferred, as tritium can diffuse through PVC bottles and equilibrate with the air.

For ^{14}C, 50–200 l samples are needed (according to the HCO_3 concentration and the laboratory requirements). These may be collected in large containers with minimum exposure to air. The detailed procedure should be discussed with the laboratory. Samples have to be delivered to the laboratory within days, to avoid equilibration with air.

Precipitation of the dissolved carbon species and their extraction from 50–200 l is preferably done in the field. This procedure requires experience (guided by the laboratory staff or, preferably, carried out by a trained hydrochemist). Precipitation in the field is time-consuming, but saves in volume and weight of samples that have to be transported to the laboratory.

For ^{13}C measurement, 3 l samples may be needed, or precipitation and extraction may be done in the field, similar to the treatment of the ^{14}C samples. Separate bottles are needed for ^{13}C, as the treatment of the large samples for ^{14}C may introduce significant fractionation in the ^{13}C value.

For noble gases samples are collected in glass tubes with special stopcocks, or in copper tubes with special clamps. The main concern is absolute avoidance of air contamination. This aspect is further discussed in Chapter 12, and should always be co-ordinated with the laboratory staff.

8.10 Preservation of samples

The storage during transport of samples from the field to the laboratory, and on the laboratory shelf, has to be done with care and expertise. Major points are:

- To keep the samples cold, as far as possible. Samples left for several hours in the sun after sampling, or samples left in a hot car for several days, may be spoiled by secondary reactions.
- To keep the samples in the dark, e.g. by placing them in boxes or storing them in dark rooms. Exposure to light enhances biological activity.
- In order to avoid biological activity, samples are occasionally poisoned by adding chemicals containing ions that will not be measured in the laboratory. Mercury iodide is a convenient preservation substance. It is available as small orange crystals, and a single small crystal added to a bottle of water is adequate. Addition of a preservative to a sample should be clearly marked on the bottle. Normally, each laboratory advocates a specific preservative.

8.11 Efflorescences

Small crystals of salts are occasionally seen to cover the soil and rocks near springs and seepages. In arid regions such efflorescences, or salt crusts, are also seen on soils in areas detached from springs or other types of surface water.

The efflorescences are formed by evaporation of rain or spring water, leaving behind the disolved salts. Infiltrating rain water redissolves such efflorescences, salinizing local groundwater. The composition of efflorescences is therefore of interest with regard to the hydrochemistry of the water they precipitated from, and with regard to the groundwater they salinize.

Tasting of efflorescences may provide immediate clues to their composition: white material with a salty taste is halite (NaCl); white insoluble material with no taste is most likely gypsum ($CaSO_4 \cdot 2H_2O$); white material that readily dissolves with a bitter taste may be epsomite ($MgSO_4 \cdot 7H_2O$), and white material that readily dissolves with no taste but a sense of cooling is most likely nitre (KNO_3 or $NaNO_3$).

Occurrence of efflorescences should be documented in the field notebook. Samples, in plastic or glass containers, should be sent to the laboratory for chemical analysis. Conceptual models developed for water systems have to incorporate the occurrence of efflorescences of observed compositions.

8.12 Equipment list for field work

A large number of items of equipment are needed for the hydrochemical fieldwork. The details vary according to the specific nature of each expedition, but experience shows that going over an equipment list is always beneficial. The following items of equipment will be required:

1. Sampling schedule (Chapter 7), maps, and writing material.
2. Field measuring equipment for temperature, conductivity, pH, dissolved oxygen and alkalinity. For these specific materials are needed for calibration, rinsing, change of batteries, or other parts.
3. Sample collection bottles. These have to be prepared carefully so that (a) all the required varieties are included (b) all bottles are clean (even new ones have to be cleaned) and (c) the bottles are all well packed in boxes – these will be needed for the filled sample bottles.
4. Filtering equipment (if filtering is to be carried out).
5. Precipitation equipment for ^{14}C and ^{13}C (if precipitation is to be done in the field).
6. Labelling material (labels or inks that have been tested to remain on the relevant bottles).

Note: Proper field measurements and well-collected samples, rapidly delivered to the laboratories, are the key to high-quality data. These are, in turn, the key to high-quality hydrochemical studies.

9 STABLE HYDROGEN AND OXYGEN ISOTOPES

9.1 Isotopic composition of water molecules

The definition of elements and isotopes was presented in sections 5.1 and 5.2. Elements are defined by the number of protons in the nucleus of the atoms. Hydrogen has one proton and oxygen has eight protons. Isotopes are defined as variations of a given element, differing from each other by the number of neutrons. As presented in section 5.2, the hydrogen isotopes are:

^1H – common hydrogen, 1 proton
^2H – deuterium (also written D), 1 proton + neutron
^3H – tritium (also written T), 1 proton + 2 neutrons

Tritium is radioactive and will be discussed in Chapter 10; common hydrogen and deuterium are stable. Oxygen has the following isotopes:

^{16}O – common oxygen, 8 protons + 8 neutrons
^{17}O – heavy (very rare) oxygen, 8 protons + 9 neutrons
^{18}O – heavy oxygen, 8 protons + 10 neutrons

Water is composed of hydrogen and oxygen, so it occurs with different isotopic combinations in its molecules. Most common and of interest to hydrochemists are: $^1H_2\ ^{16}O$ (common), $^1HD\ ^{16}O$ (rare), and $^1H_2\ ^{18}O$ (rare). The water molecules may be divided into light molecules ($^1H_2\ ^{16}O$) and heavy water molecules ($^1HD\ ^{16}O$ and $^1H_2\ ^{18}O$).

9.2 Units of isotopic composition of water

The isotopic composition of water is expressed in comparison to the isotopic composition of ocean water. For this purpose an internationally agreed sample of ocean water has been selected, called Standard Mean Ocean Water, or in brief: SMOW (Craig, 1961a, b).

The isotopic composition of water, determined by mass spectrometry, is expressed in per mil (‰) deviations from the SMOW standard. These deviations are written δD for the deuterium, and δ¹⁸O for ¹⁸O:

$$\delta D\text{‰} = \frac{(D/H)_{sample} - (D/H)_{SMOW}}{(D/H)_{SMOW}} \times 1000$$

and

$$\delta^{18}O\text{‰} = \frac{(^{18}O/^{16}O)_{sample} - (^{18}O/^{16}O)_{SMOW}}{(^{18}O/^{16}O)_{SMOW}} \times 1000$$

Water with less deuterium than SMOW has a negative δD, and water with more deuterium than SMOW has a positive δD. The same is true for δ¹⁸O.

The isotopic composition of water is measured by mass spectrometry. A 50 ml water sample is ample.

9.3 Isotopic fractionation during evaporation and some hydrological applications

Evaporation is a physical process in which energy-loaded water molecules move from the water phase into the vapour phase. Isotopically light water molecules evaporate more efficiently than the heavy ones. As a result, an isotopic fractionation occurs at partial evaporation of water: the vapour is enriched in light water molecules, reflected in relatively negative δD and δ¹⁸O values. In contrast, the residual water phase becomes relatively enriched in the heavy isotopes, reflected in more positive δD and δ¹⁸O values. The isotopic separation, or fractionation, is more efficient if the

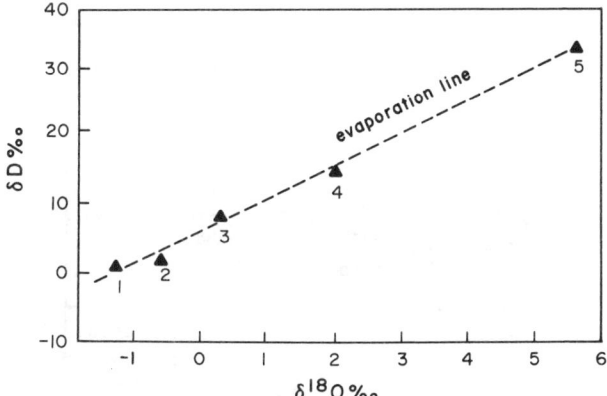

Fig. 9.1 Isotopic composition of water samples successively collected at an evaporating pond at Qatar (Yurtsever and Payne, 1979). The water in the pond became progressively heavier in its composition.

produced vapour is constantly removed, e.g. by wind blowing away vapours produced above an evaporating pond. Figure 9.1 presents the composition of water samples successively collected at a pond in Qatar (Yurtsever and Payne, 1979). The composition of the samples is plotted on a δD-$\delta^{18}O$ diagram. A progressive enrichment in the heavy isotopes is noticed, indicating that the residual water in the pond was progressively enriched in δD and $\delta^{18}O$ as the isotopically light vapour was removed. The original water had a composition of $\delta D = 0‰$, $\delta^{18}O = -1.4‰$ and the last sample reached values of $\delta D = +33‰$, $\delta^{18}O = +5.6‰$. The line connecting the sample points is called the evaporation line, and its slope is determined primarily by the prevailing temperature and air humidity.

In surface water bodies that evaporate to a very great extent, the isotopic enrichment ceases or is even reversed. However, except in very rare cases, natural water bodies, such as lakes or rivers, are only partially evaporated, in the range in which isotopic enrichment of the residual water occurs.

Isotopic fractionation during evaporation causes fractionation during cloud formation: the vapour in the clouds has a lighter isotopic composition than the ocean that supplied the water. Upon condensation from the cloud, during rain formation, the reverse is true: the heavy water molecules condense more efficiently, leaving the cloud residual vapour depleted of deuterium and oxygen-18.

Residual evaporation water is thus tagged by high δD and $\delta D^{18}O$ values, an observation used to trace mixing of evaporation brines with local fresh water. An example of two mineral springs located on the Dead Sea shore is given in Fig. 9.2. The Hamei Zohar and Hamei Yesha springs are seen to lie on a fresh water – Dead Sea mixing line, indicating that these springs contain recycled Dead Sea water, brought to the surface with the emerging fresh water, recharged at the Judean Mountains (Gat et al., 1969).

From the information included in Fig 9.2, one can calculate the percentage of Dead Sea brine intermixed in the Hamei Yesha and the Hamei Zohar springs.

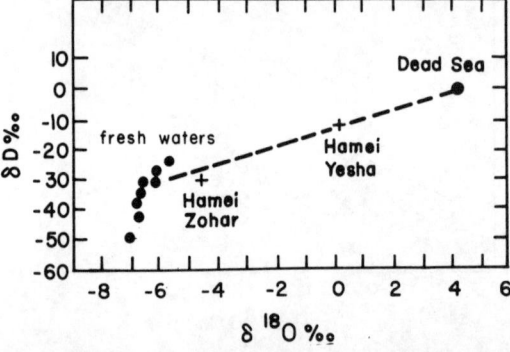

Fig. 9.2 Stable hydrogen and oxygen isotopic composition of the Hamei Zohar and Hamei Yesha mineral springs, Dead Sea shores. The linear correlation indicates that the springs' water is formed by intermixing of Dead Sea brine with local fresh water (Gat et al., 1969).

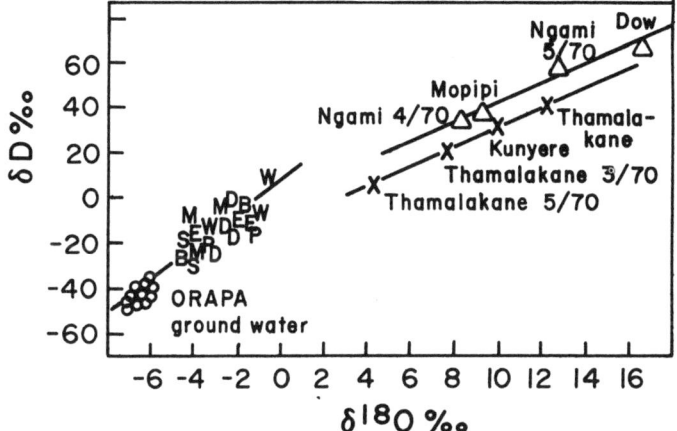

Fig. 9.3 Isotopic data of confined groundwater at Orapa, Botswana (O), average annual rain in neighbouring meteorological stations (letters) and 45 km distant lakes (Δ) and rivers (×) (Mazor et al., 1977).

At Orapa, northern Botswana, a new well field of confined water was developed and two hypotheses were proposed to explain the origin of recharge: either underground flow of recharge water from lakes and rivers 45 km distant, or local rain. The consultants of the local diamond mine ruled out local recharge, and favoured replenishment from the lakes. Results of an isotopic composition survey depict in Fig. 9.3 the δD and $\delta D^{18}O$ values of the groundwater in the wells, the average annual rain composition in neighbouring meteorological stations, and the lakes and rivers mentioned. The δD and $\delta^{18}O$ values of the groundwater in the wells were observed to be significantly lighter (more negative) than the lake and river waters, which were enriched in the heavy isotopes due to intensive evaporation losses. Thus, the hypothesis of recharge from the lakes could be ruled out, and recharge by local rain was supported. However, a slight, but analytically significant, difference can be seen between the composition of average annual rain and local groundwater. This was explained by an observation that only intensive rains are effectively recharging the groundwater, and these were observed by Vogel and Van Urk (1978) to have an isotopic composition that is lighter than average annual rain composition (the amount effect, section 9.6).

9.4 The meteoric isotope line

Harmon Craig published (1961a) a $\delta D - \delta^{18}O$ diagram, based on about 400 water samples of rivers, lakes, and precipitation from various countries (Fig. 9.4). An impressive lining of the data along the best-fit line of

$$\delta D = \delta^{18}O + 10$$

has been obtained. Outside this line plot data from East African lakes that undergo significant isotopic fractionation due to intensive evaporation losses.

The data in Fig. 9.4 lie on a straight line in spite of the very wide range of values: δD of $-300‰$ to $+50‰$, and $\delta^{18}O$ of $46‰$ to $+6‰$. This line, called the *meteoric line*, has been found, with some local variations, to be valid over large parts of the world.

The meteoric line is a convenient reference line for the understanding and tracing of local groundwater origins and movements. Hence, in each hydrochemical investigation the *local meteoric line* has to be established from samples of individual rain events, or monthly means of precipitation. A specific example of a local meteoric line, from north-eastern Brazil, is given in Fig. 9.5. A local meteoric line is obtained from the equation $\delta D = 6.4 \, \delta^{18}O + 5.5$ (Salati et al., 1980). Examples of equations of local meteoric lines, reported from various parts of the world, are given in Table 9.1.

The composition of precipitation is reflected, directly or modified, in the composition of groundwater. Common practice is to plot groundwater data on $\delta D - \delta^{18}O$ diagrams, along with the meteoric line of local precipitation as a reference line. Examples are given in Figs. 9.6-9.8. In Fig. 9.6 the composition of the groundwater plots are close to the local meteoric line, ruling out secondary processes, such as evaporation prior infiltration, or isotope exchange with aquifer rocks.

Table 9.1 Examples of regional meteoric lines

Region	Meteoric line (‰)	Reference
'Global' meteoric line	$\delta D = 8\delta^{18}O + 10$	Craig (1961)
Northern hemisphere, continental	$\delta D = (8.1 \pm 1) \delta^{18}O + (11 \pm 1)$	Dansgaard (1964)
Mediterranean or Middle East	$\delta D = 8\delta^{18}O + 22$	Gat (1971)
Maritime Alps (April 1976)	$\delta D = (8.0 \pm 0.1) \delta^{18}O + (12.1 \pm 1.3)$	Bortolami et al. (1979)
Maritime Alps (October 1976)	$\delta D = (7.9 \pm 0.2) \delta^{18}O + (13.4 \pm 2.6)$	
North eastern Brazil	$\delta D = 6.4 \, \delta^{18}O + 5.5$	Salati et al. (1980)
Northern Chile	$\delta D = 7.9 \, \delta^{18}O + 9.5$	Fritz et al. (1979)
Tropical islands	$\delta D = (4.6 \pm 0.4) \delta^{18}O + (0.1 \pm 1.6)$	Dansgaard (1964)

In Fig. 9.7 the groundwater data fall distinctly below the relevant meteoric line, indicating that secondary fractionation has occurred, or that the waters are ancient and were recharged in a different climatic regime.

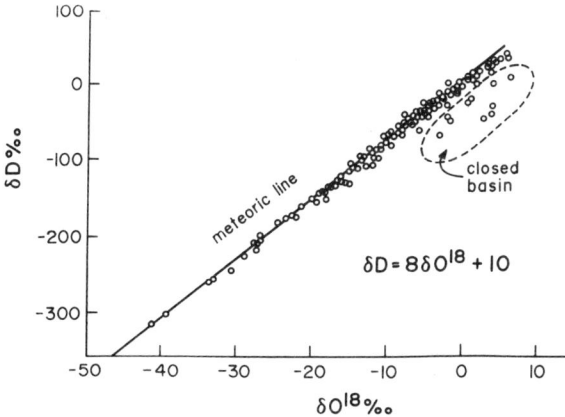

Fig. 9.4 Isotopic data of about 400 samples of rivers, lakes and precipitation from various parts of the world. The best-fit line was termed the *meteoric line*. Its equation, as found by Craig (1961a), is $\delta D = 8\delta^{18}O + 10$. The data in the encircled zone of 'closed basins' is for East African lakes with intensive evaporation.

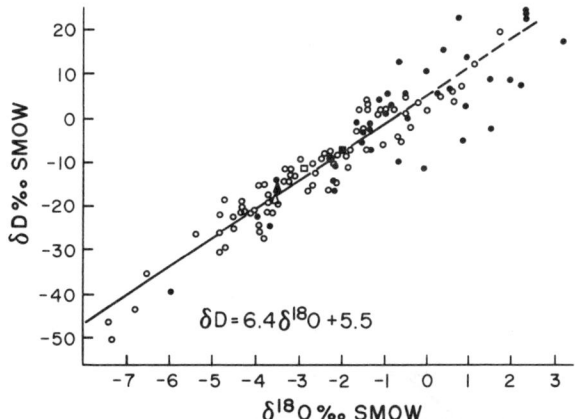

Fig. 9.5 Isotopic composition of precipitation in the Pajeu River Basin, Brazil. o, months with rain over 50 mm/month; •, months with lower precipitation amounts. A local meteoric line is obtained with the equation $\delta D = 6.4\delta^{18}O + 5.5$ (Salati *et al.*, 1980).

What should be checked in order to decide between these two possible explanations? The age. Groundwater data plotted in Fig. 9.8 fall along the meteoric line but reveal a large spread of values. Separation into shallow and deep waters revealed the latter have distinct lighter isotopic compositions. A possible origin from ancient recharge, at a different climatic regime, was suggested in this case too. The necessary check has been done: the deeper waters, of light isotopic composition, were indeed found to contain little ^{14}C, indicating high groundwater ages (section 11.4), supporting the paleoclimatic hypothesis.

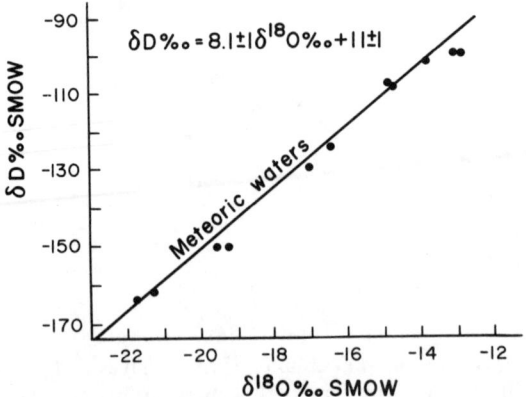

Fig. 9.6 Isotopic composition of water sampled from wells in central Manitoba, Canada. The values fall close to the local meteoric line ($\delta D = 8.1\delta^{18}O + 11$). Hence, the researchers (Fritz *et al.*, 1974) concluded that evaporation during recharge, and isotopic exchange with aquifer rocks, are insignificant.

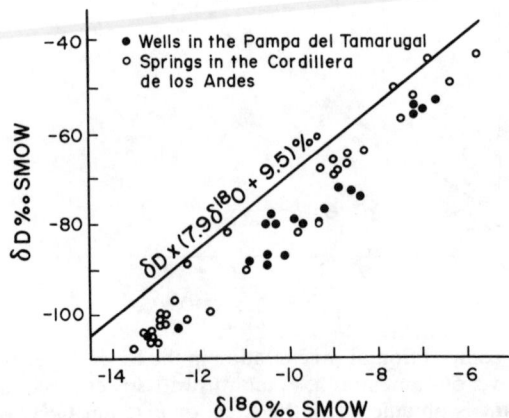

Fig. 9.7 Isotopic composition of groundwaters of nothern Chile. The values lie below the meteoric line of local precipitation, explained by the investigators (Fritz *et al.*, 1979) as reflecting secondary fractionation by evaporation prior to infiltration, or presence of ancient waters that originated in a different climatic regime. The large variations in the groundwater compositions are useful in local groundwater tracing.

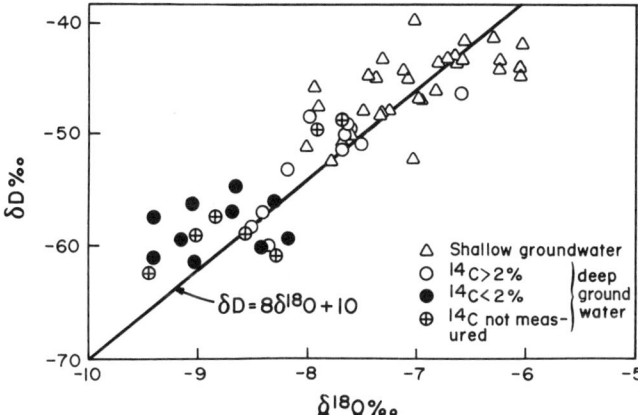

Fig. 9.8 Isotopic composition of groundwaters near Chatt-el-Honda, Algeria (Gonfiantini *et al.*, 1974). The values scatter along the meteoric line but reveal an internal order: values of deep groundwaters are isotopically lighter (more negative) than shallow groundwaters. This was taken as an indication that the deep groundwaters were ancient and originated from rains of a different climatic regime. This is supported by the large number of ancient deep groundwater water samples, as borne out by low ^{14}C concentrations (sections 11.4 and 11.8).

9.5 Temperature effect

Dansgaard (1964) analysed a large body of isotopic data gathered by the International Atomic Energy Agency, and showed that temperature is the major parameter that determines the isotopic values of precipitation (Fig. 9.9). In his extensive discussion, Dansgaard summed up the knowledge gained in laboratory experiments and field observations. The composition of precipitation depends on the temperature at which the oceanic water is evaporated into the air and, even more important, the temperature of condensation at which clouds and rain or snow are formed. The net effect is expressed in the following empirical function (from Fig. 9.9):

$$\delta^{18}O = 0.7\ T_a - 13‰, \text{ or: } 0.7‰/°C$$

and, in a similar way:

$$\delta D = 5.6\ T_a - 100‰, \text{ or } 5.6‰/°C$$

A local study at Heidelberg, Germany, revealed the following empirical function (Schoch-Fischer *et al.*, 1983):

$$\delta D = (3.1\ \pm 0.2)\ T_a - (172\ \pm 3)‰, \text{ or: } (2.8\ \pm 0.2)‰/°C$$

(T_a is the local mean annual air temperature.) The meteoric line (Fig. 9.4) is, thus, the result of the combined δD and $\delta^{18}O$ dependences on temperature. The temperature effect is well seen in seasonal variations in regions with rains during cold and warm seasons. An elegant example from Switzerland

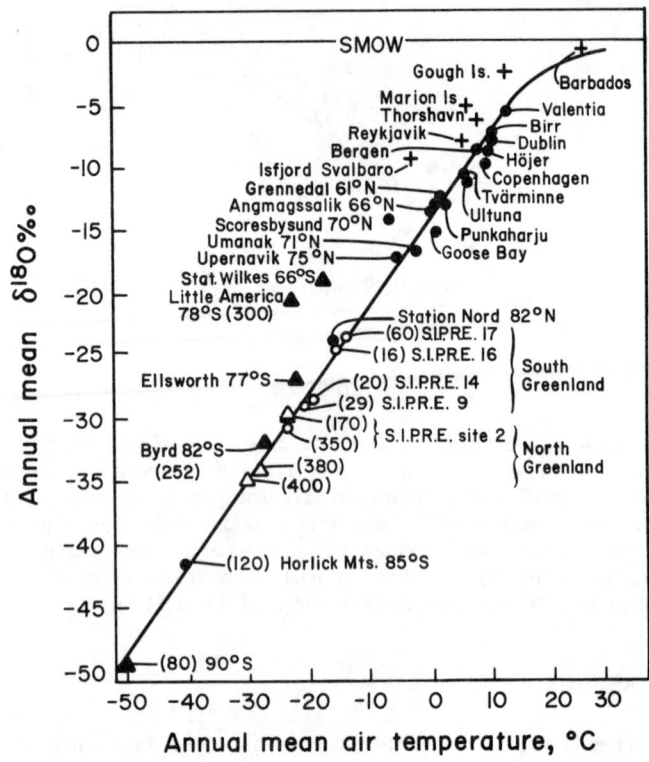

Fig. 9.9 Temperature effect. Correlation between annual mean $\delta^{18}O$ values observed in precipitation and annual mean temperature of local air. Polar ice (circles and triangles; figures in parenthesis indicate total thickness in cm); continental precipitation (•) and island precipitation (+) (Dansgaard, 1964).

is given in Fig. 9.10: monthly measurements in three stations revealed an annual temperature cycle, followed by corresponding changes of $\delta^{18}O$.

The temperature dependence of the isotopic composition of precipitation, or *temperature effect*, is to a large extent responsible for the large variation in the isotopic composition of groundwaters, thus equipping the hydrologist with a powerful tool. In regions with summer and winter precipitation the isotopic differences in the composition of the precipitation are tracable as winter recharge and summer recharge fronts, important in establishing groundwater velocities and identifying piston flows.

9.6 Amount effect

Figures 9.11–9.13 show the dependence of the isotopic composition of the amount of rain: the heavier the rain event, or the greater the amount of

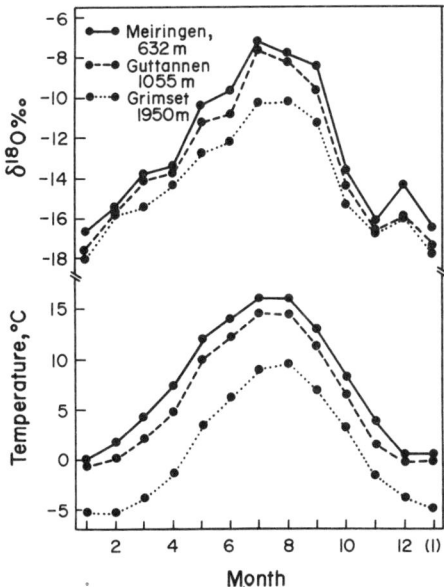

Fig. 9.10 Monthly mean $\delta^{18}O$ values in precipitation, and monthly mean air temperatures, during 1971-8, for Swiss stations. The value of January (1) is shown twice to complete the cycle (Siegenthaler and Oeschger, 1980). The $\delta^{18}O$ values are seen to co-vary with the temperature, reflecting a pronounced temperature effect of 0.35-0.5‰ $\delta^{18}O/°C$. The measurements, carried out at three stations of different altitudes, revealed an altitude effect, precipitation at higher altitudes having isotopically lighter compositions.

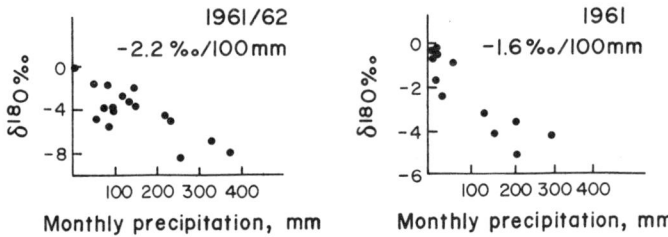

Fig. 9.11 Two cases of isotopic compositions varying with the amount of precipitation (the amount effect), reported by Dansgaard (1964). Left, Binza (Leopoldville), Congo; right, Wake Island. The $\delta^{18}O$ precipitation values are of individual months. The amount effects were -2.2‰ $\delta^{18}O/100$ mm rain for Binza and -1.6‰ $\delta D/100$ mm rain for Wake Island.

monthly precipitation, the more negative are the δD and $\delta^{18}O$ values. Dansgaard (1964) proposed two major explanations for this *amount effect*:

- Lower ambient temperatures cause the formation of clouds with lighter isotopic composition (temperature effect, Fig. 9.9) and lower temperatures also cause heavier rains.

- Falling rain drops undergo evaporation, enriching the falling rain in the heavy isotopes. This effect is less severe both when ambient temperatures are low, and when the amount of rain is large (as the air gets more saturated).

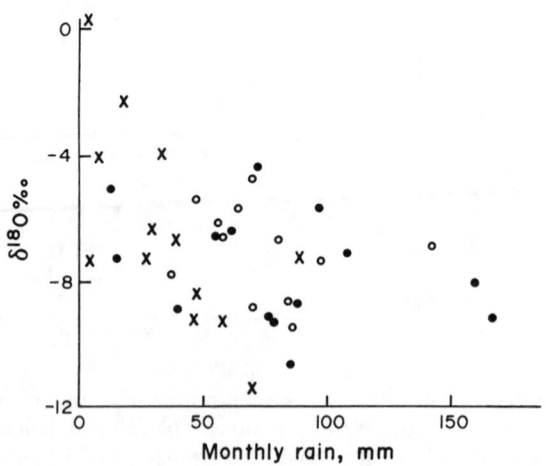

Fig. 9.12 Amount effect. Monthly rain and $\delta^{18}O$ values measured for three years at Rowen Boos, Haute Normandie, France. Dots, 10/74 to 12/75; ×, 1976, , 1977 (after Conrad et al., 1979). Rainier months reveal isotopically lighter rain.

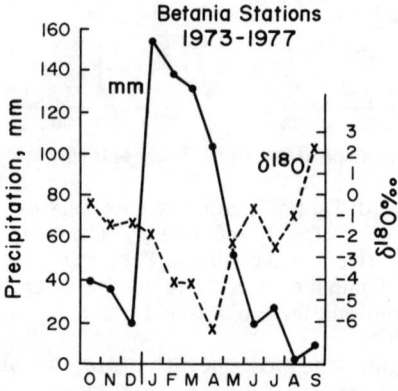

Fig. 9.13 Amount effect. Monthly rain and $\delta^{18}O$ values of north-eastern Brazil plotted as a function of sampling date. The curves are mirror shaped, revealing low $\delta^{18}O$ values at the high rain months (Salati et al., 1980).

The amount of monthly rain varies in most cases during the year, causing a seasonal variation in the isotopic composition. This point is demonstrated in a case study from north-eastern Brazil (Fig. 9.13).

9.7 Continental effect

Several workers noticed that the average isotopic composition of precipitation tends to have more negative values further away from the ocean coast. This *continental effect* is well reflected in European groundwaters (Fig. 9.14). The explanation lies in the history of the precipitating air masses. As they travel inland, rain is gradually precipitated by condensation, accompanied by more efficient condensation of water molecules with heavier isotopes (opposite to evaporation). The *residual moisture* in the air masses thus becomes gradually lighter in its isotopic composition, and lighter rain is progressively formed. The continental effect is often masked by other effects, e.g. temperature (seasonal) effect and altitude effect, as discussed in the next section.

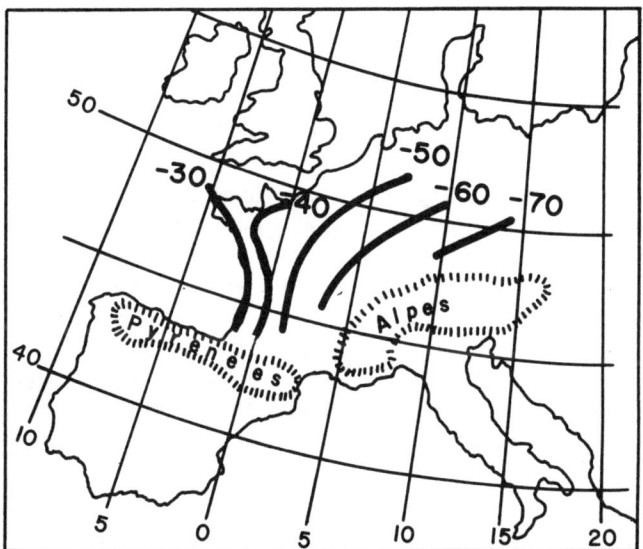

Fig. 9.14 Lines of equal δD values for Europe, based on over 300 samples of groundwater. A trend of lighter isotopic composition is seen as a function of the distance from the ocean, reflecting the continental effect in the precipitation (after Sonntag *et al.*, 1979).

9.8 Altitude effect

An altitude effect is seen in Fig. 9.15. The $\delta^{18}O$ values in Swiss precipitation are lighter with higher altitudes. The gradient, or *altitude effect*, is

$-0.26‰\ \delta^{18}O/100$ m altitude

The altitude effect is observed in this case to be the same in precipitation and in the derived surface and groundwaters. This effect is not masked by the seasonal temperature effect (Fig. 9.10).

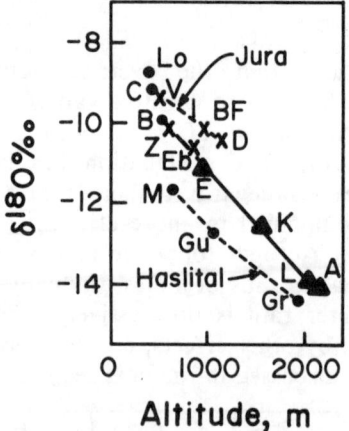

Fig. 9.15 Variations of mean $\delta^{18}O$ as a function of altitude for Swiss precipitation (•), groundwaters (×) and rivers (▲). The lines are parallel, all revealing an altitude effect of $-0.26‰$ $\delta^{18}O/100$ m (Siegenthaler and Oeschger, 1980).

The altitude effect has to be established in each study area. Leontiadis et al. (1983) reported a value of $-0.44‰/100$ m for the part of Greece that borders Bulgaria.

As a cloud rises up the mountains, the heavy isotopes are depleted and the residual precipitation gets isotopically lighter. This effect turns out to be an effective tool in tracing groundwater recharge.

Payne and Yurtsever (1974) reported a case study in which isotope hydrology was recruited to find out to what extent groundwaters in the Chinahdega Plain, Nicaragua (1100 km², 200 masl) were recharged by local precipitation on the plain, or by recharge from the higher slopes of the Cordillera Mountains (up to 1745 m). They worked in three stages:

- They analysed $\delta^{18}O$ in weighted mean precipitation samples from different altitudes, and defined an altitude effect of $-0.26‰/100$ m (Fig. 9.16).
- They analysed groundwaters with known recharge altitudes, and found the same altitude effect (Fig. 9.17).
- They measured the $\delta^{18}O$ of the studied waters in the plain and established the ratios of local plain recharge (around 25%) to mountain recharge.

A nice control was tritium, found to be higher in wells with higher plain-recharge contributions, revealing shorter travel times, as compared to low tritium contents in wells with higher contributions of mountain recharge (longer travel times).

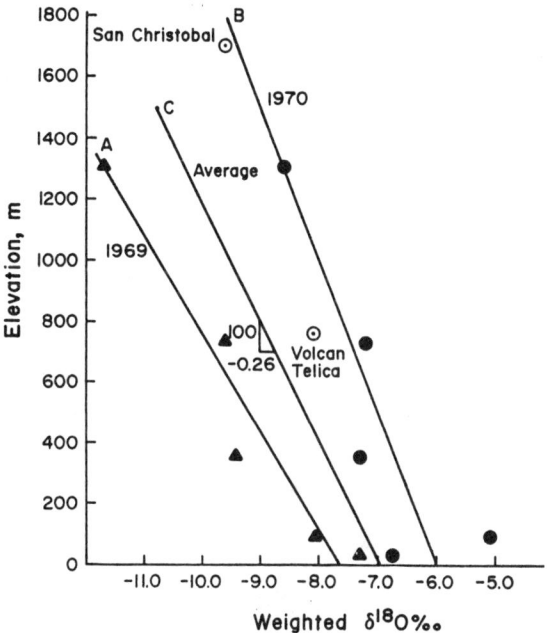

Fig. 9.16 $\delta^{18}O$ values in mean weighted samples of precipitation from different altitudes in Nicaragua. The lines for 1969 (A) and for 1970 (B) are quite parallel, the average (line C) revealing an average altitude effect gradient of $-0.26‰$ $\delta^{18}O/100$ m (Payne and Yurtsever, 1974).

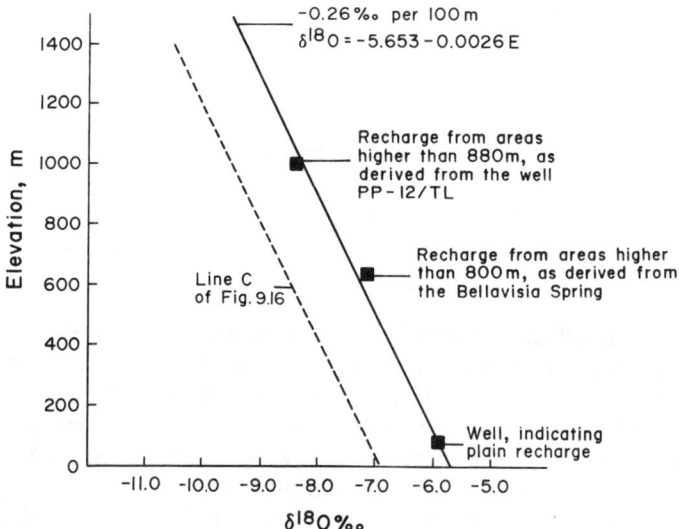

Fig. 9.17 Altitude effect, reflected in groundwater of known recharge areas, Nicaragua. An effect of $-0.26‰$ $\delta^{18}O/100$ m is obtained, the same as observed for the regional precipitation (Fig. 9.16). (Payne and Yurtsever, 1974).

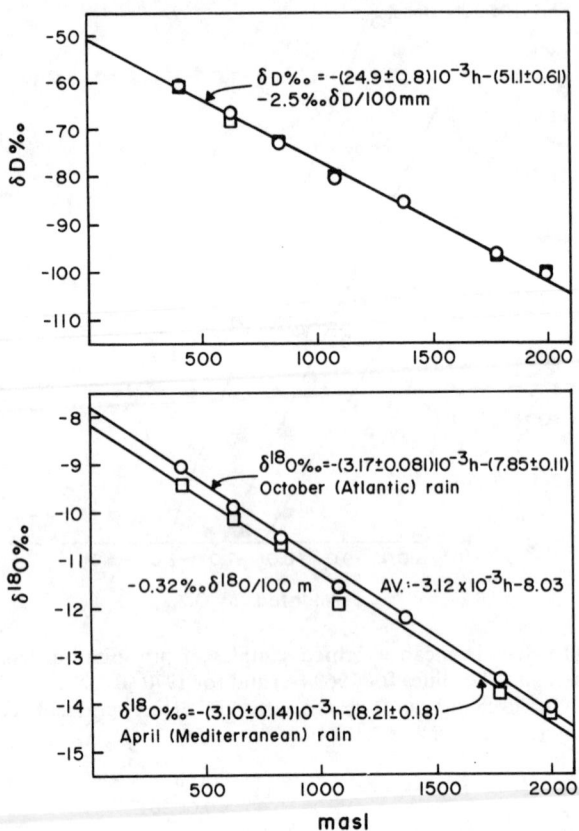

Fig. 9.18 Isotopic compositions as a function of altitude for precipitation samples collected in a summer month (April 1976) and a winter month (October 1974), Maritime Alps. The data were used to establish local altitude effect equations (see text). Winter and summer $\delta^{18}O$ values fall on seperate lines, reflecting the difference in the origin of the precipitating air masses – from the Atlantic and Mediterranean respectively (after Bortolami et al., 1978).

The altitude effect is applied in an ever-growing number of studies as a tool to calculate the recharge altitude from the D and $\delta^{18}O$ data of springs and wells. Hence, much interest lies in a study of Bortolami et al. (1978) who checked the accuracy of such calculations in a case study in the Italian Maritime Alps. They conducted the following measurements: δD and $\delta^{18}O$ in precipitation in stations of various altitudes, during one winter month (April 1976) and one summer month (October 1974). The data (Fig. 9.18) revealed different lines for the winter and summer $\delta^{18}O$ values, and only one line for δD. The corresponding altitude–isotope equations were found from Fig. 9.18 to be:

$$\delta^{18}O = -3.12 \times 10^{-3}h - 8.03$$

and

$$\delta D = 24.9 \times 10^{-3}h - 51.1$$

where h is the altitude in metres above sea level (masl). Accordingly, two altitude equations may be written.

$$h = -320 \, \delta^{18}O - 2564$$

and

$$h = -40.2 \, \delta D - 2052$$

In the second stage, samples were repeatedly collected for isotopic analysis in the karstic Bossea Cave, at an altitude of 810 masl. The data, given in Table 9.2, were used individually to calculate recharge altitudes. The agreement between hD and $h^{18}O$ is good in the various samples, but the agreement between repeated samples is rather poor. Values of calculated recharge altitude vary from 1180 m to 1871 m. This is a highly discouraging range of values, showing that calculations based on single water samples are non-reliable. The *average hD and $h^{18}O$ values*, 1565 m and 1539 m, seem, however, to be very reasonable as average recharge altitudes for the water sampled in the Bossea Cave. The calculations based on the *average δD and $\delta^{18}O$ values* are practically identical to the earlier ones: 1563 and 1538 m.

The research of Bortolami and his co-workers (1979) provides an excellent example of a study aimed at a local calibration of an isotopic tracing technique. The variability observed between results of single

Table 9.2 Recharge altitude calculated from δD and $\delta^{18}O$ in repeatedly collected samples in the Bossea Cave, Maritime Alps, Italy (810 masl)[a]

Data	δD‰	$\delta^{18}O$‰	hD(m)	$h^{18}O$(m)
13-10-74	−81.4	−11.70	1220	1180
7-3-76	−88.4	−12.86	1502	1551
9-4-76	−94.4	−12.86	1743	1551
18-4-76	−97.3	−13.83	1859	1861
25-4-76	−97.6	−13.85	1871	1868
1-5-76	−90.2	−12.91	1574	1567
14-5-76	−85.8	−12.34	1397	1385
30-5-76	−87.6	−12.54	1470	1449
6-6-76	−87.5	−12.51	1465	1440
Average	−89.93	−12.82	1565[b]	1539[b]
			1563[c]	1538[c]

[a] Data from Bortolami *et al.* (1979) applying their average equation for the local precipitation:

$h = 320 \, \delta^{18}O - 2564$ and $h = -40.2 \, \delta D - 2052$ (text).

[b] Average of recharge altitudes calculated via δD and $\delta^{18}O$.
[c] Calculation based on average $\delta D = -89.9$‰ and average $\delta^{18}O = -12.82$‰.

samplings is probably extreme, as they worked on a karstic system which has a rapid response to individual storm events.

9.9 Tracing groundwater with deuterium and oxygen-18: local studies

In work on South African warm springs, the question arose whether their temperature and dissolved ions represent the water at depth, or did intermixing with shallow cold water take place, especially from adjacent rivers. Hence, δD and $\delta^{18}O$ were measured in the thermal springs and the nearest rivers (Mazor and Verhagen, 1983). The results are shown in Fig. 9.19. It is seen that the thermal springs have significantly more negative values than the nearby rivers, indicating that no intermixing took place except in one case. This conclusion has been supported by the practical absence of measurable tritium in the thermal springs, indicating long travel times. What might be the reason for the lighter isotopic composition of the South African warm springs, as compared to the river waters, as seen in Fig. 9.19? Three explanations can be offered:

- An altitude effect, caused by spring recharge in higher elevations. This would agree with the increased temperatures, indicating relatively deep circulation.
- The waters might be ancient, belonging to a different climatic regime (paleowaters).
- The river waters were heavy at the time of sampling, in the context of the seasonal variations.

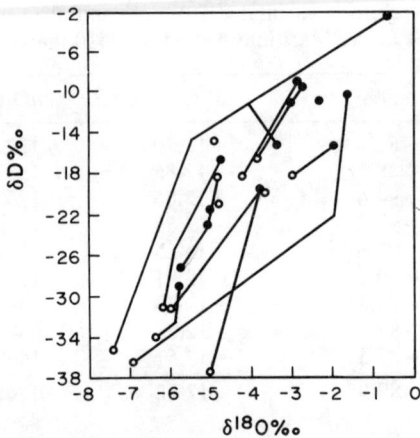

Fig. 9.19 Isotopic composition of thermal waters (o) and adjacent rivers (•) in South Africa. The lines connect thermal springs with the nearest river. It is seen that the spring waters are significantly lighter than the nearby rivers, indicating that no (or negligible) intermixing occurred (Mazor and Verhagen, 1983).

The uncertainty regarding the causes of the isotopic differences did not intervene in the application of this difference to a tracing problem (non-intermixing of thermal and surface waters).

Rivers and shallow groundwater are often saline in arid or semi-arid regions. Examples are common in north-eastern Brazil in crystalline rocks, remote from the sea, and in regions where no marine transgression has occurred for 100 million years. Hence, evaporation was for a long time suspected to be the cause of the rise in salinity. The problem was studied by examining Cl and $\delta^{18}O$ (Fig. 9.20). In general, the Cl and $\delta^{18}O$ curves co-varied during the months of observation, proving that evaporation caused the salinity (Salati et al., 1980). A variance is seen in the details of the peaks of the two curves in Fig. 9.20, e.g. the peak defined by the last three points. This might be explained by special rain events, e.g. a summer rain that underwent much evaporation during descent to the ground (resulting in more positive $\delta^{18}O$), which lowered for a short while the Cl concentration in the river water (at the low-flow season).

In shallow groundwaters, at a depth of about 1 m, in the Chott-el-Honda salt plain of inland Algeria, a positive correlation was observed between dry residue and $\delta^{18}O$ (Fig. 9.21). Evaporation through the thin soil cover was deduced (Gonfiantini et al., 1974). This conclustion has been confirmed by the more saline water lying on an evaporation line in a δD-$\delta^{18}O$ diagram (Fig. 9.22).

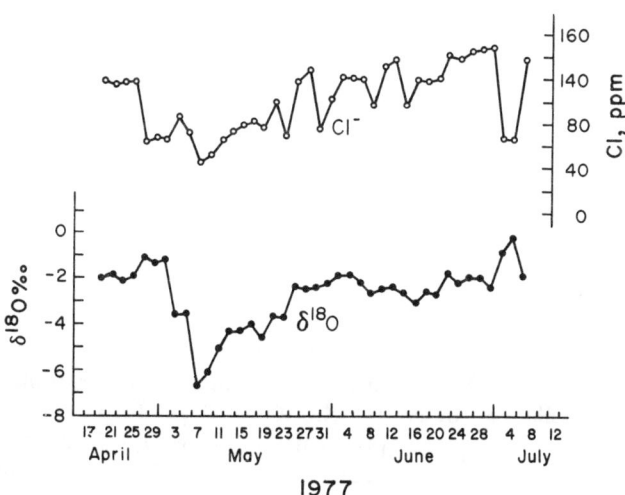

Fig. 9.20 Cloride and $\delta^{18}O$ values in repeatedly collected samples of the Pajeu River, north-eastern Brazil. A general correlation is seen, revealing the role of evaporation, most important during July (rise in ambient temperatures and low river flow) (Salati et al., 1980).

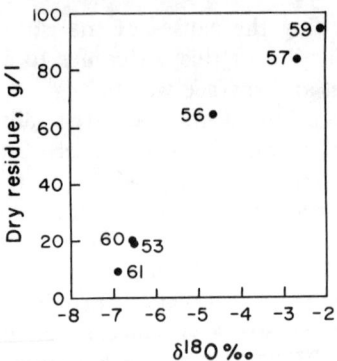

Fig. 9.21 Isotopic composition and salinity (dry residue) in shallow wells south of the Chott-el-Honda salt plain, Algeria. Evaporation is evident, but a further check can be seen in the next figure (Gonfiantini *et al.*, 1974).

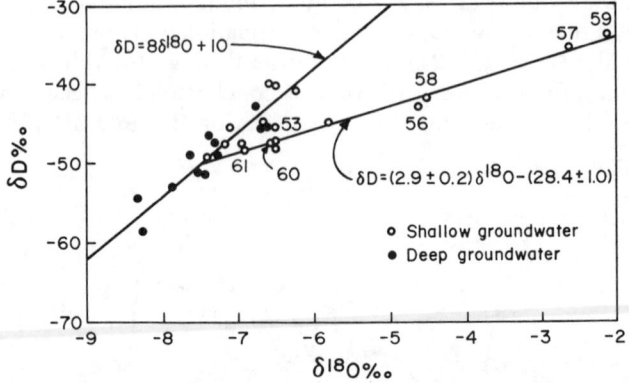

Fig. 9.22 Isotopic composition of groundwaters south of Chott-el-Honda, Algeria. The shallow groundwaters lie on an evaporation line and the degree of heavy isotope enrichment agrees with the salinity (the sample numbers are the same as in the previous figure) (Gonfiantini *et al.*, 1974).

The stable isotopes of water are most useful in tracing sea water intrusions, so troublesome in coastal urban areas. One example is seen in Fig. 9.23 for wells west of Hermosillo, Gulf of California.

Table 9.3 reports data on depth profiles in coastal wells of the Salentine peninsula, Italy. Which trends can be seen and what is their meaning in terms of groundwater movement? In general, temperature, Cl, δD, $\delta^{18}O$ and $\delta^{13}C$ values are high in the deeper parts of the wells, reflecting sea water encroachment. The temperature gradient is caused by the regional geothermal gradient. Practical hydrological conclusions may be reached with regard to desired pumping management of these wells, i.e. filling up lower parts of wells, and/or limited pumping rates of fresh water only.

Table 9.3 Water depth profiles in coastal wells in the Salentine peninsula, Italy (data extracted from Cotecchia et al., 1974)

No.	Source	Date	°C	Cl⁻ (meq/l)	HCO₃ (meq/l)	$\delta^{18}O$(‰)	δD(‰)	$\delta^{13}C$(‰)	^{14}C(pmc)
1	Boraco Spring	9-2-71	17.4	29.2	5.94	−5.5	−34.5	−10.70	41.1
2	Chidro Spring	3-2-71	18.4	59.8	6.15	−5.3	−33.5	−9.50	24.3
3	Iduse Spring	14-7-71	18.2	114.8	4.29	−4.2	−26.5	−9.54	52.7
4	CH well, 40 m	2-2-71	17.4	39.5	6.23	−5.9	−33.5	−10.80	35.1
5	69 m	2-2-71	17.5	42.0	6.24	−5.7	−33.0	−11.30	49.8
6	86 m	3-2-71	17.5	54.9	6.20	−5.7	−32.0	−10.70	35.7
7	118 m	2-2-71	17.5	55.2	6.12	−5.6	−32.0	−12.10	30.4
8	147 m	30-7-71	17.4	594.9	2.40	+1.8	+10.0	−7.09	—
9	170 m	30-7-71	18.0	594.3	2.53	+1.7	+11.0	−5.17	5.5
10	S-1 well, 19 m	9-7-71	17.0	31.9	5.24	−5.4	−31.5	−11.84	64.0
11	30 m	9-7-71	17.0	44.0	5.63	−5.3	−31.0	−13.40	65.4
12	40 m	13-7-71	17.1	581.9	2.36	+1.3	+8.5	−5.00	46.7
13	S3 well, 9 m	13-7-71	17.2	55.4	5.48	−5.1	−29.5	−13.93	66.9
14	19 m	13-7-71	17.4	580.2	2.40	+1.3	+8.5	−3.70	92.6
15	C-S well, 56 m	23-2-71	16.8	2.3	4.37	−5.1	−31.0	−14.37	77.2
16	108 m	24-2-71	16.8	37.8	6.00	−4.9	−30.5	−11.30	—
17	175 m	25-2-71	17.1	562.3	2.68	+1.0	+6.3	−8.87	1.4
18	184 m	9-12-72	17.1	565.2	—	+0.8	+6.1	−7.60	—
19	30 m	17-2-71	17.2	7.3	4.93	−6.10	−36.1	−14.40	64.2
20	70 m	18-2-71	17.5	12.3	4.79	−5.9	−34.0	−13.32	55.8
21	SR well, 146 m	19-2-71	19.4	587.3	2.65	+1.2	+7.5	−5.61	1.8
22	154 m	5-12-72	19.4	594.3	—	+1.4	+9.2	−5.29	—
23	C-S well, 33 m	14-7-71	16.6	15.4	4.59	−5.0	−33.5	−10.90	51.4
24	70 m	14-7-71	18.4	581.5	2.46	+1.3	+9.0	−4.09	31.7
28	Ionian Sea	13-7-71	—	593.2	2.42	+1.3	+9.0	−3.30	111.9
29	Adriatic Sea	14-10-73	—	583.4	2.25	+0.8	+7.5	−2.60	81.0

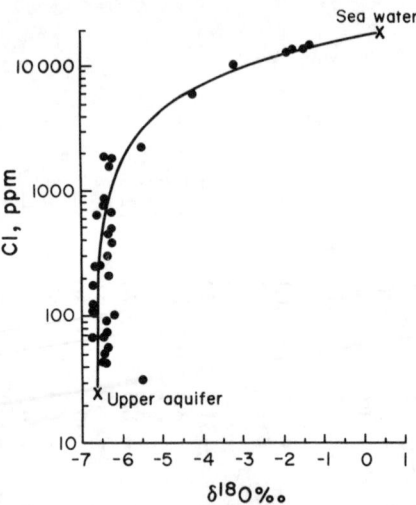

Fig. 9.23 Chlorinity and isotopic composition for coastal wells, Hermosillo, Gulf of California. The percentage sea water infiltration can be calculated for each well (Payne *et al.*, 1980). A mixing line is seen, its curvature being caused by the use of a logarithmic chloride axis.

9.10 Tracing groundwater with D and oxygen-18: a regional study

Stahl *et al.* (1974) provided an excellent example of isotopic tracing of groundwater. In the Sperkhios Valley, Greece (insert in Fig. 9.24). The research followed the following logical steps:

- Artesian wells were grouped according to their δD and $\delta^{18}O$ values and geographical distribution. Three such groups emerged (Table 9.4 and Fig. 9.25).
- Springs for which the recharge altitude could be deduced from field data were analysed and an isotope-altitude graph was established, defining the local altitude effect (Fig. 9.26).
- The recharge altitudes of the three artesian well groups were determined with the aid of the altitude–isotopic composition graph (Fig. 9.26), calibrated by the springs of known intake altitudes. The values are given in Table. 9.4.
- Once the three well groups and their corresponding recharge altitudes were obtained, the details of local flow were worked out by additional chemical, radioisotope and field data (Fig. 9.24).
- A group of saline (2.5 to 14.5 g Cl/l) and warm (28–40°C) springs and wells, in the eastern part of the study area, were suspected by earlier workers to contain a sea water component. This was tested and confirmed by Cl–$\delta^{18}O$ relations, and the percentage of sea water intermixed in each case could be calculated (Fig. 9.27).

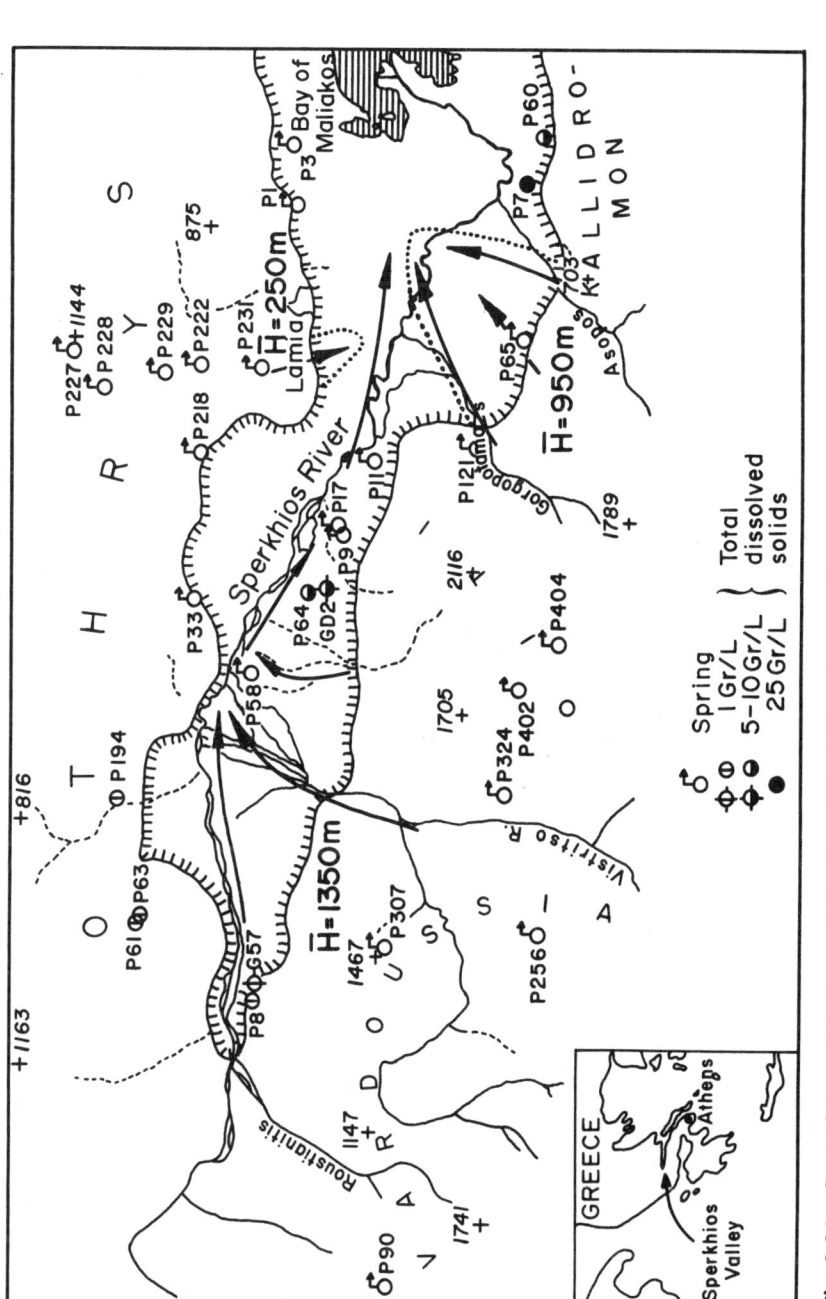

Fig. 9.24 General map of study area in the Sperkhios Valley, Greece. The recharge altitudes (H) were calculated from D and ^{18}O data (Stahl et al., 1974).

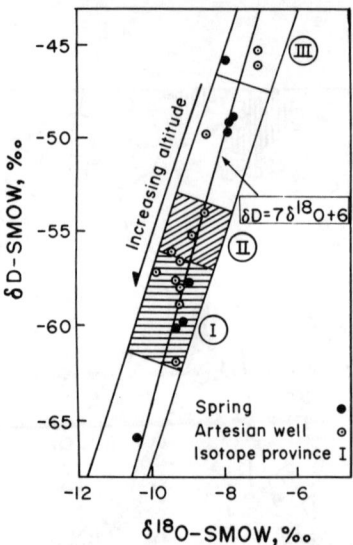

Fig. 9.25 Isotopic data of artesian wells and springs in the Sperkhios Valley. Three well groups, or isotopic provinces, were recognized. A local meteoric line, $\delta D = 7\delta^{18}O + 6$, was established from the spring data (Stahl et al., 1974).

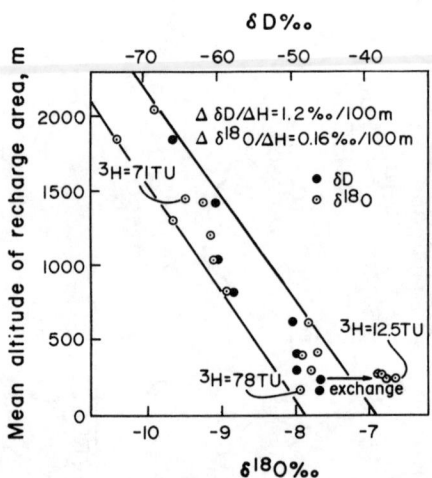

Fig. 9.26 Isotopic composition and recharge altitude of springs in the Sperkhios Valley (Stahl et al., 1974).

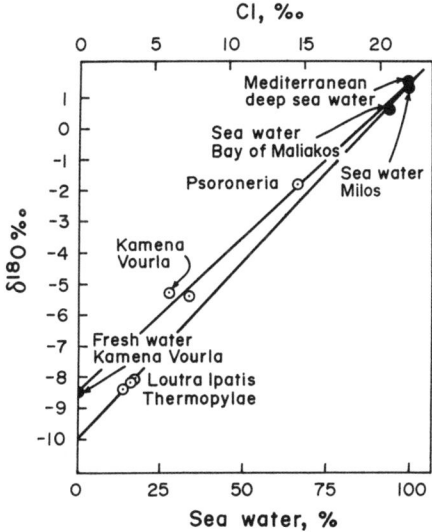

Fig. 9.27 Isotopic composition and chlorinity of warm springs at the bay of Maliakos (Fig. 6.24). Values lie on a mixing line with sea water, proving that the last one is intermixing in the mineral springs (Stahl et al., 1974).

Table 9.4 Isotopically deduced average recharge altitudes for Artesian well groups in the Sperkhios Valley (following data by Stahl et al., 1974)

Well group	Av. $\delta^{18}O$(‰)	Av. recharge altitude (masl)
I	-9.36 ± 0.04	1350
II	-8.81 ± 0.07	950
III	-7.80 ± 0.15	250

9.11 The need for multisampling

In the previous sections the many factors that determine the isotopic composition of groundwater have been described. Many of these, especially the temperature, vary seasonally and others, e.g. intensity of rain-evaporation, vary even from one rain event to the other. As a result, single samples may be non-representative. Table 9.5 gives examples for the variations between daily precipitations in a single station.

Significant variations in the isotopic composition were seen in water samples repeatedly collected in a pond (Fig. 9.1), rains samples collected at different dates (Fig. 9.5), water samples from adjacent wells (Fig. 9.6), samples from wells of different depths (Fig. 9.8 and Table 9.3), precipitation samples of various months (Fig. 9.10), rains of different intensities (Fig. 9.11), summer and winter precipitation (Fig. 9.18), or river samples collected during a whole season (Fig. 9.20).

The basic conclusion from these variations is the necessity of *multisampling*. The hydrochemist has to plan the sample collection in the right way in order to get enough data for the calculation of meaningful average values and to gain an insight into the fine structure of the investigated system in the lateral, vertical and time dimensions.

Note. An excellent collection of basic isotope data and discussion of case studies is provided by publications of the International Atomic Energy Agency, Vienna, included in the reference list on p. 270. Especially useful is the collection of papers *Stable Isotope Hydrology: Deuterium and Oxygen-18 in the Water Cycle*, IAEA, Vienna, 1981. Fritz and Fontes (1980 and 1986) are the editors of the *Handbook of Environmental Isotope Geochemistry* that contains valuable information on the stable hydrogen and oxygen isotopes.

Table 9.5 Variations in $\delta^{18}O$ values in daily rain at Ashdod, Israel (Gat and Dansgaard, 1972)

Date (Jan 1968)	Amount (mm)	$\delta^{18}O$ (‰ SMOW)
6	5.3	5.01 ± 0.06
9	6.5	2.09 ± 0.06
10	1.0	0.14 ± 0.07
14	12.5	6.13 ± 0.07
15	10.1	6.43 ± 0.13
16	5.2	2.05 ± 0.13
17	3.3	2.3 ± 0.11
21	3.4	4.70 ± 0.11
23	2.4	2.70 ± 0.10
24	13.0	5.27 ± 0.11
28	7.1	3.09 ± 0.11
30	10.7	4.15 ± 0.11
31	7.6	6.90 ± 0.11

10 TRITIUM

10.1 The radioactive heavy hydrogen isotope

Tritium is the heavy isotope of hydrogen (section 5.2). Its symbol is ^3H, or T. Tritium atoms are unstable and disintegrate radioactively, forming stable ^3He atoms. The radioactive decay is accompanied by the emission of β^- particles, measurable in specific laboratories:

$$T \xrightarrow[12.3y]{} -\beta + {}^3He$$

The rate of radioactive decay is by convention expressed as the half-life, $T_{1/2}$, defined as the time span during which a given concentration of the radioelement atoms decays to half the initial value. $T_{1/2}$ of tritium is 12.3 years. Thus, after 12.3 years ½ the initial concentration of tritium atoms is left, after 24.6 years only ¼ is left and so on. A radioactive decay curve of tritium is given in Fig. 10.1. Using the decay curve it is possible to find out, for example, how many years it takes for a given amount of tritium to decay to 20% of the initial amount. The answer, obtained from Fig. 10.1, is 29 years. Similarly, one can find out what percentage of an initial amount of tritium will be left after 20 years. The answer is 32% (Fig. 10.1).

The concentration of tritium in water is expressed by the ratio of T atoms to H atoms:

a ratio of $T/H = 10^{-18}$ is defined as **1 tritium unit (1 TU)**

10.2 Natural tritium production

Cosmic ray neutrons interact in the upper atmosphere with nitrogen, producing ^{15}N, which is radioactive and disintegrates into common carbon (^{12}C) and tritium:

$$^{14}N + n \rightarrow {}^{15}N \rightarrow {}^{12}C + {}^3H$$

The tritium atoms are oxidized to water and become mixed with precipitation, and so enter groundwater. The natural production of tritium introduces about 5 TU to precipitation and surface water.

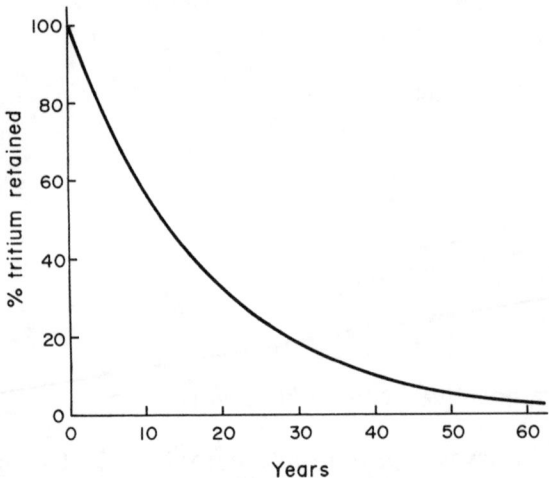

Fig. 10.1 Radioactive decay curve of tritium. After 12.3 y 50% of an initial concentration are left, after 24.6 y 25% are left, etc.

In the saturated zone, water is isolated from the atmosphere and the tritium concentration drops due to radioactive decay: the original tritium concentration of 5 TU drops to 2.5 TU after 12.3 years, only 1.2 TU are left after another 12.3 years and so on.

Provided the natural production of 5 TU is the only source of tritium in groundwater, we would have at hand a water dating tool. For example, water pumped from a well with 3 TU has preserved $3 \times 100/5 = 60\%$ of its natural tritium content, equivalent (Fig. 10.1) to an age of 9 years.

However, a hydrologist's life is not that simple. Most aquifers have a capacity equal to several annual recharges or, in other words, water accumulates in the aquifer for many years and the age we have just calculated from the triutium concentration is not a real age, but rather an average, or *effective age*. This is still of high value in hydrology. Tritium measurements have a practical limit of 0.2 TU. Hence, assuming an initial input of 5 TU, an amount of 0.2 TU represents $0.2 \times 100/5 = 4\%$, or an age of 55 years. Thus, the tritium method of effective hydrological age seems valid in the range of a few decades.

Low tritium concentrations are determined on samples that have been pre-concentrated electrolytically, necessitating samples of about 300 ml. Thus, to allow for repeated measurements, samples of 1 l should be collected for tritium measurements.

10.3 Manmade tritium inputs

Hydrogen bomb tests, which began in 1952 in the northern hemisphere, added large amounts of tritium to the atmosphere. They reached a peak in

1963, with up to 10 000 TU in a single monthly rain in the USA. An international treaty stopped surface nuclear tests in 1963, and concentrations in precipitation decreased steadily. Since nuclear testing began tritium (and δD and $\delta^{18}O$) are measured in a worldwide net of stations (Fig. 10.2), co-ordinated by the International Atomic Energy Agency (IAEA) in Vienna. The results are published in annual reports: *Environmental Isotope Data; World Survey of Isotope Concentration in Precipitation*, IAEA, Vienna. Figure 10.3 represents annual tritium concentrations in precipitation at various stations between 1961 and 1975. Figure 10.4 reveals monthly values for several stations over the period of 1961-1965, when the bomb-tritium impact was especially high. The following patterns are seen in the tritium curves:

- Values rose much higher in the northern hemisphere, where the bomb testing took place.
- The maximum peak was reached in 1963, and values decreased thereafter.
- A summer peak and winter low are seen each year (Figs. 10.4 and 10.5), reflecting the fact that the tritium was displaced in large amounts in the higher parts of the atmosphere and leaked in the spring and summer into the lower parts.

The manmade tritium, reaching several thousand TU in precipitation during 1963, completely masked the natural tritium production discussed in the previous section. The awareness of the potential importance of tritium to hydrology arose only after the nuclear tests began. By that time the natural tritium content in precipitation could no longer be measured, but a unique solution was found – measurements in stored and dated wine bottles, reflecting the relevant annual rains (Table 10.1). A common value

Table 10.1 Estimates of natural, pre-bomb, tritum concentration in precipitation and wine (text) (Vogel et al., 1974)

Locality	Sample	TU	Researchers[a]
Rhone valley, France	wine	3.4	Kaufman and Libby (1954) Von Buttlar and Libby (1955)
Bordeaux, France	wine	4.3	Kaufman and Libby (1954) Von Buttlar and Libby (1955)
Rhine valley, Germany	wine	5.5	Roether (1967)
New York State, USA	wine	5.8	Kaufman and Libby (1954) Von Buttlar and Libby (1955)
Chicago, USA	wine	7.5	Kaufman and Libby (1954) Von Buttlar and Libby (1955)
Ottawa, Canada	rain	15.3	Brown and Gummit (1956) Brown (1961)
Greenland	ice	12.6	Begemann (1961)

[a] References given in Vogel et al. (1974).

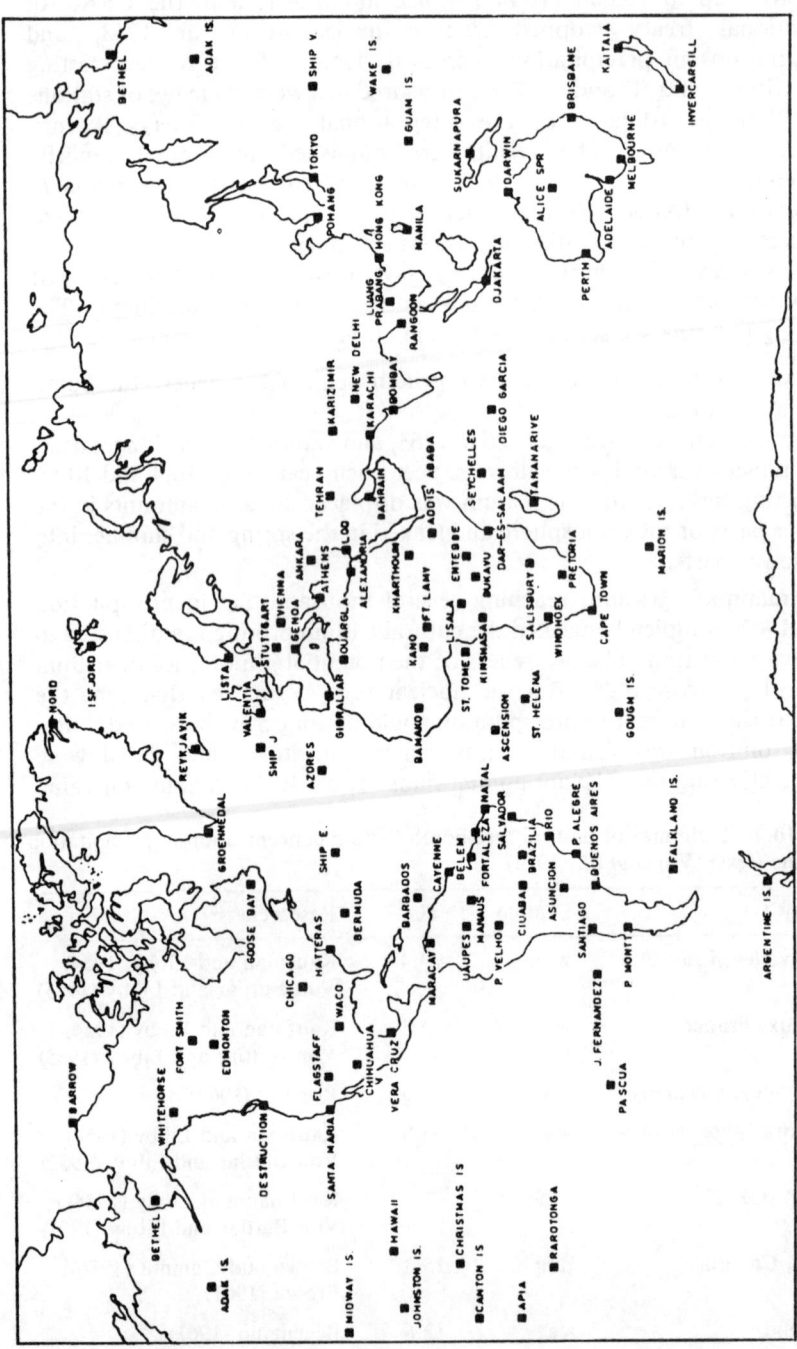

Fig. 10.2 Network of weather stations collecting precipitation samples for tritium and stable hydrogen and oxygen measurements, co-ordinated by the International Atomic Energy Agency, Vienna.

Tritium

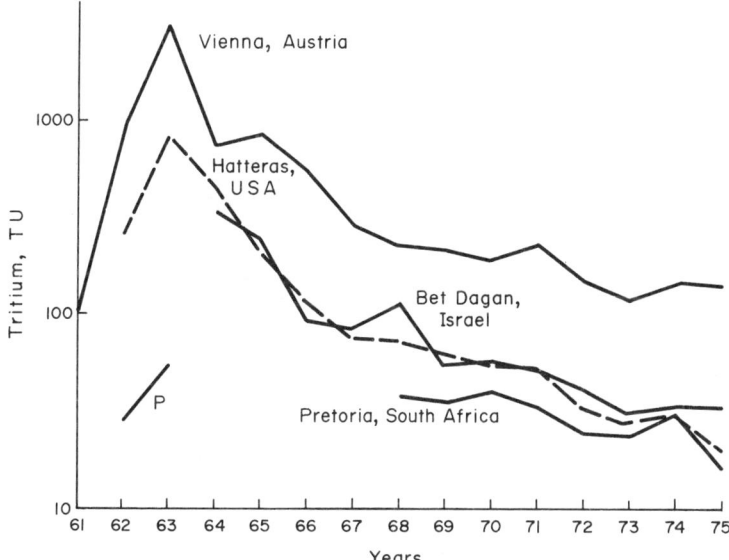

Fig. 10.3 Annual weighted average tritium concentrations of representative weather stations (data from Environmental Isotope Data: World Survey of Isotope Concentrations in Precipitation, Vol.1 (1969) and Vol.7 (1983)). Values increased drastically in 1963 and have been steadily decreasing ever since. Values were much higher in the northern hemisphere (where the nuclear tests took place).

for pre-bomb tritium is 5 TU.

The described observations may be applied to practical problems. For example, the weighted average tritium concentration in Vienna was 3280 TU in 1963 (Fig. 10.3). How much tritium would be left in water that was recharged in 1963 and pumped in a well in 1988? The answer may be attained by the following steps:

- 1988−1963 = 25 years passed, i.e.
- 25% of the inital tritium were left after decay (read from Fig. 10.1), and
- 3280 × 25/100 = 820 TU would be left at 1988.

Before 1952, natural tritium was about 5 TU in precipitation. How much would be left in groundwater recharged in 1951 and emerging in a spring in 1988? The answer may be reached by the following steps: 37 years passed, therefore radioactive decay would have decreased the concentration to 11% of the initial value (Fig. 10.1), so 5 × 11/100 = 0.5 TU would be left.

10.4 Tritium as a short-term age indicator

When the anthropogenic tritium was noticed, hopes arose that the specific tritium 'pulses' contributed by the individual tests would provide a means of accurate dating of groundwater. However, it turned out that the input

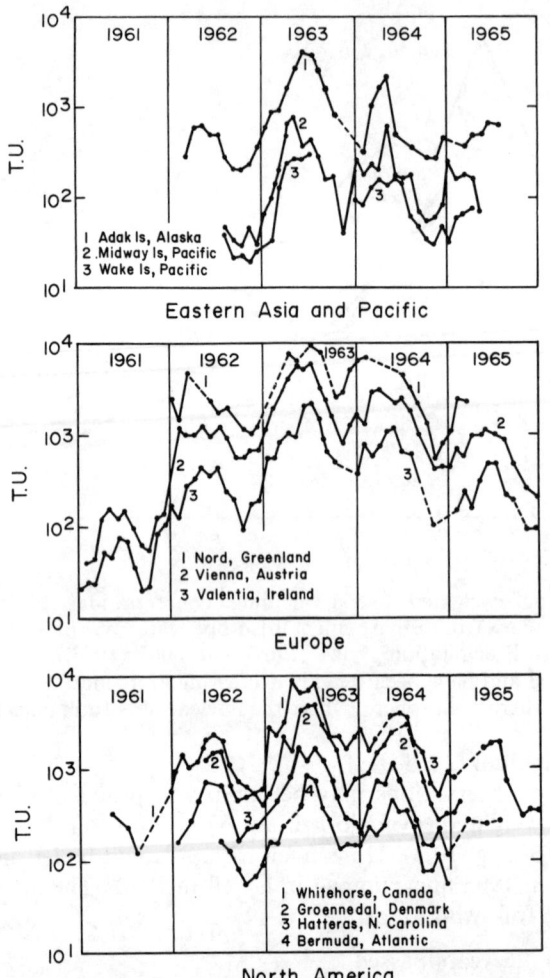

Fig. 10.4 Monthly tritium concentrations in rain at representative stations (from IAEA Technical Report No. 73, 1967). Summer peaks indicate that bomb tritium is stored in the high parts of the atmosphere and leaks in the spring into the lower atmosphere.

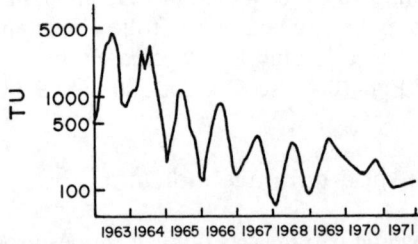

Fig. 10.5 Summer peaks and winter lows of tritium in precipitation in Sweden (Pearson, 1974).

values in precipitation varied considerably from one location to another, and during the seasons and years in each place. In addition, complicated mixing takes place in each aquifer, and the mode and extent of mixing of each year's recharge with that of previous years' recharge is unknown. Hence, age determinations accurate to the year are impossible and of no meaning to groundwater studies. However, semi-quantitative dating is possible and very informative:

- Water with zero tritium (in practice <0.5 TU) has a pre-1952 age.
- Water with significant tritium concentrations (in practice > 10 TU) is of a post-1952 age.
- Water with little, but measurable, tritium (between 0.5 and 10 TU) seems to be a mixture of pre-1952 and post-1952 water. The topic of water mixtures is further discussed in section 11.10, along with carbon-14 data.

10.5 Tritium as a tracer of recharge and piston flow: observations in wells

Seasonal variations in the tritium concentration in two shallow tribal wells in the Kalahari are seen in Fig. 10.6. Variations in tritium were noticed to follow variations in the water table. This observation provided insight into the local recharge mechanism. The tritium-rich recharge of the rainy season formed an upper layer in the aquifer. As abstraction, mainly during the dry season, advanced, lower water layers were encountered, having been stored in the aquifer for several years. The seasonal tritium peak was observed to arrive at the wells several months after the rain peak. Thus, the recharge water moved in the aerated zone not through conduits, but in a porous medium, in a piston flow mode (section 2.1, Fig. 2.2). The time lag provided an idea of the storage capacity of the aerated zone.

Tritium in soil profiles. Several workers measured tritium concentrations in profiles of soil moisture. The samples were collected in most cases by means of a hand drill; the profile samples were carefully wrapped to avoid drying

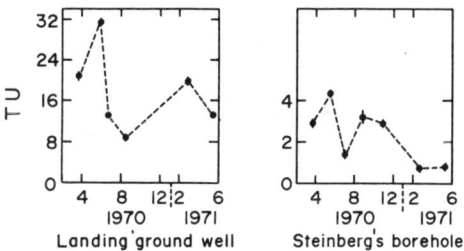

Fig. 10.6 Seasonal variations in tritium concentration in shallow wells in the Kalahari, indicating recharge in a piston-flow mode (Mazor *et al.*, 1973).

or exchange with tritium in the air. In the laboratory the soil moisture was extracted by distillation, weighted (to get a moisture profile) and measured for the tritium concentration. Figure 10.7 deals with such a profile from South Africa, along with the local precipitation-tritium curve. The profile was taken in December 1971 and the precipitation-tritium curve has been corrected (solid line, top of Fig. 10.7) for radioactive decay until that date. This corrected line indicates 7 TU for 1958 precipitation, an increase to over 20 TU in the rainy season of 1962, and over 30TU for 1964. These values were applied to identify the time points marked on the tritium-depth line in Fig. 10.7. Thus, the moisture with the 1958 tritium concentration was observed in 1971 at a depth of about 2 m. The infiltrating water moved these 2 m in 1971 − 1958 = 13 years, indicating an average infiltration velocity of 0.15 m/y. The 1962 tritium concentration was observed at a depth of 1.4 m, after 1971 − 1962 = 9 years, indicating infiltration velocity of 0.16 m/y; and the 1964 tritium front was observed at 0.8 m, indicating an infiltration velocity of 0.11 m/y. Variations in the calculated infiltration velocity may occur either due to differences in the soil properties or due to variations in the annual rain regimes. One may therefore retrieve from soil profiles average recharge velocities, relative lateral and vertical conductivity (by comparing different profiles), and recharge efficiencies.

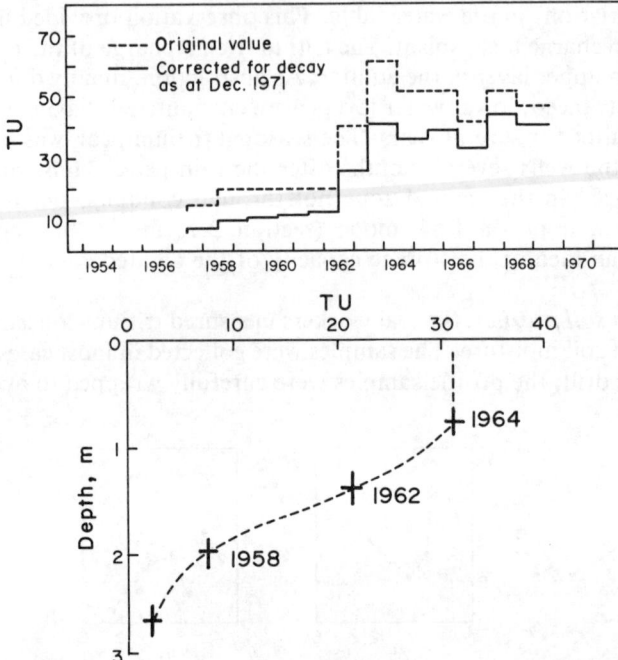

Fig. 10.7 Tritium in a soil profile from the Transvaal, South Africa, and in precipitation at Pretoria (following Bredenkamp *et al.*, 1974). The identification of the 1958, 1962, and 1964 moisture fronts is discussed in the text, along with the applications to recharge percentage calculations.

Applicativity of tritium data in soil profiles. The applicability of the tritium profile data may be demonstrated by means of an example: a tritium front in a soil profile was identified to have descended 3 m during 10 years and the weighted average moisture content in this section was found to be 12% (by volume). These data may serve to calculate the average annual recharge: out of the 3 m profile, $3 \times 0.12 = 0.36$ m were the equivalent column of recharged water. Hence, the average annual recharge was $0.36/10 = 0.036$ m, or 36 mm per year. Continuing with this example, if the average annual precipitation was 650 mm/y, then the average recharge percentage was $36 \times 100/650 = 5.5\%$.

A second example for the application of tritium data: a tritium concentration signifying the 1962 peak was observed at a depth of 4.8 m in a soil profile with a weighted average moisture content of 12% by volume. The observation was made in 1978, and the local average precipitation was 720 mm/y. What was the average infiltration velocity and percentage? The average infiltration velocity was $4.8/16 = 0.3$ m/y, the equivalent recharge annual water column was $0.3 \times 0.12 = 0.036$ m, or 36 mm, and hence the percentage recharge was $36 \times 100/720 = 5\%$.

Active rain recharge in a desert indicated in a tritium soil profile. Figure 10.8 portrays tritium profiles in soil moisture near Grootfontein, South Africa (Bredenkamp *et al.*, 1974). A layered structure is easily seen, indicating a piston flow type of recharge in the soil. Another set of tritium soil profiles from the Southern Kalahari, South Africa, is given in Fig. 10.9 (Verhagen *et al.*, 1979). The value of 10 TU was adopted to indicate the arrival of the 1963 tritium front. The results of this study are of special interest in the light of a long-standing controversy about whether recharge is effective at all in the sand-covered Kalahari dry land (Mazor, 1982). The 1963 recharge has penetrated 7–23 m, thus demonstrating active recharge. Profile B.H. 3 was taken below a large thorn tree. The depth of recharge penetration was only 8 m, about half the depth in the adjacent profiles. This was attributed to the water consumption of the tree.

Soil profiles ignore water flow in conduits. The tritium profiles were interpreted with the basic premise that the soil is a uniform medium, with a sponge-like structure, through which water infiltrates. However, short-circuit flow through conduits, such as cracks, bioturbations, or decayed root channels, must also be considered. Cracks or rodent holes are efficient intake points of runoff, and occasionally become filled with coarse particles, providing routes of relatively high conductivity. When a soil profile builds up, these coarsely repacked channels of higher conductivity add up, forming a network of higher conductivity conduits. The soil cores will not sample water flowing in cracks or burrowed holes. Thus, the infiltration velocities and percentages, calculated by the above discussed tritium–soil profiles, should be regarded as *minimum* values.

The importance of short-circuit infiltration may vary significantly from

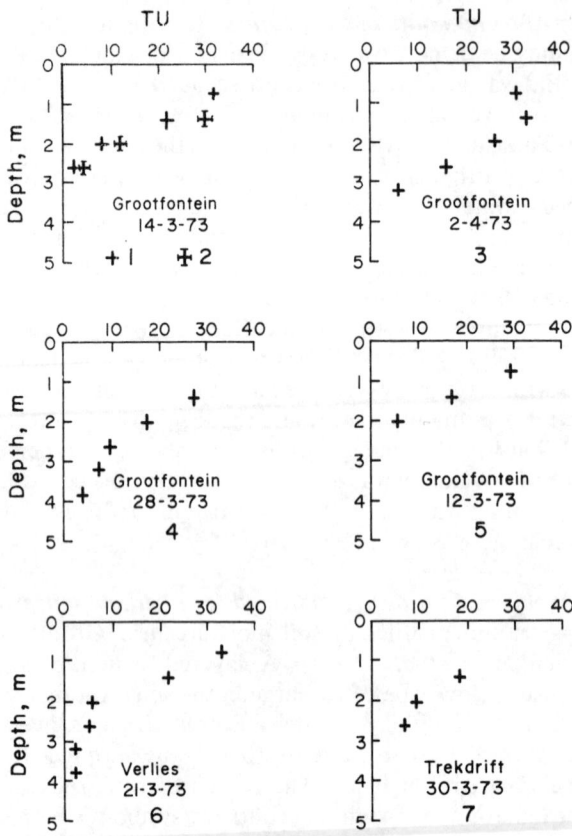

Fig. 10.8 Variations of tritium concentration with depth in soil profiles near Grootfontein, South Africa (from Bredenkamp *et al.*, 1974). Such tritium profiles are an excellent means of demonstrating piston flow of infiltrating recharge water.

one area to another and its intensity may be measured by tritium measurements, e.g. in dripping water in shallow mines or in groundwater in flat terrains: if, in such studies, tritium values encountered are higher than in the overlaying soil prefiles, then short-circuit recharge has occurred. Such studies are meaningful in cases where lateral recharge can be excluded on grounds of local geology, or topography (e.g. flat terrain or hill tops).

Two case studies. Figure 10.10 gives data on repeated tritium measurements in a well and in the adjacent Mohawk river. What hydrological conclusions may be drawn? The variations in tritium concentrations in the well followed variations in the river, as shown by letters in Fig. 10.10. Hence, the river is

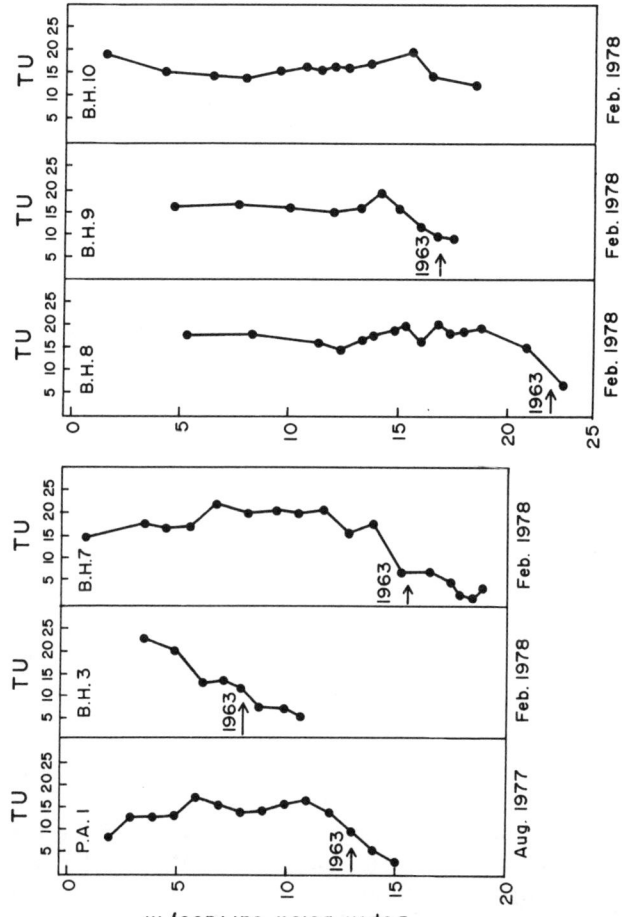

Fig. 10.9 Tritium in soil moisture profiles in the southern Kalahari, South Africa (from Verhagen et al., 1979). The measurements were done in 1978, and 10 TU was taken to mark the 1963 recharge. B.H. 3 was profiled under a large camel thorn tree.

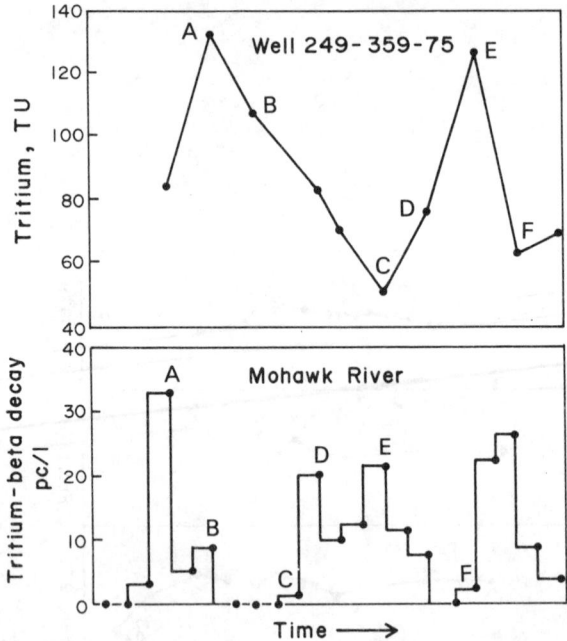

Fig. 10.10 Repeated tritium measurements in a well and in the adjacent Mohawk River, New York, USA (from Winslow et al., 1965). The river was monitored for tritium by beta decay counting and the results are expressed in pc/l (10^{-12} curie/l). Letters mark tritium levels in the river, identified in the well. Conclusions: (a) The well responded with a brief time lag to changes in the river, hence it is recharged by the river. (b) The recharge flows underground in a piston flow (orderly arrival of water 'fronts').

recharging the well. The time lag in the well's response may be used to calculate recharge velocities. The data reveal piston flow of the recharge water, with little smoothening by dispersion (Fig. 10.10). Tritium may be used for regional recharge studies. An example from the Qatar Peninsula has been reported by Yurtsever and Payne (1979). They measured tritium in a large number of wells in a shallow aquifer, producing a map with iso-tritium contours (Fig. 10.11). Recharge occurred in areas of high tritium, flowing to areas of low tritium contents. The reader is referred to a paper by Allison and Hughes (1975), discussing the use of tritium measurements to estimate amount of recharge.

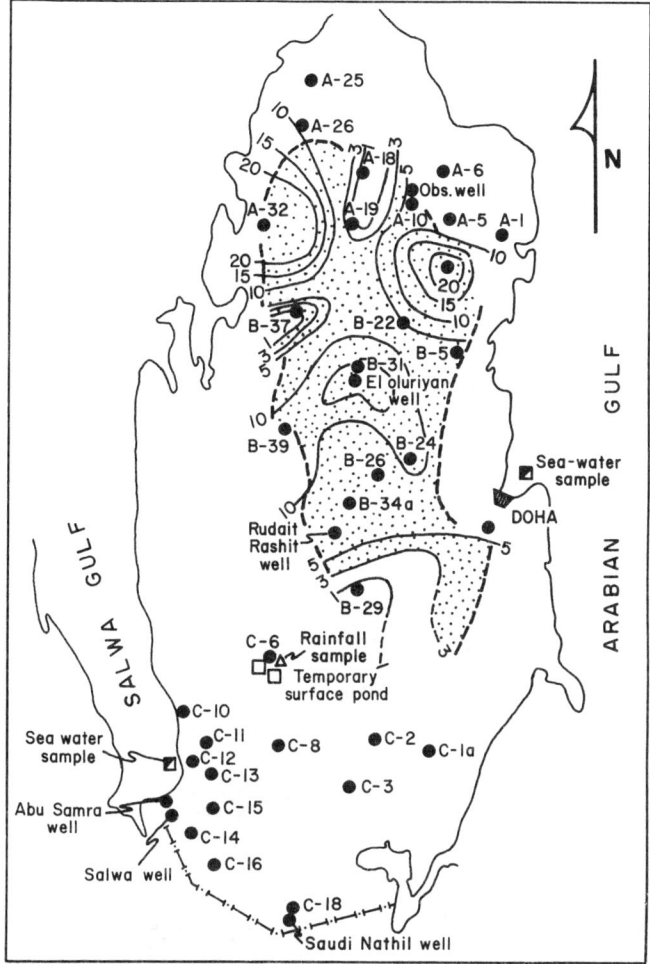

Fig. 10.11 Iso-tritium lines (in TU) for wells in a shallow aquifer on the Qatar peninsula (from Yurtsever and Payne, 1979). Recharge is from areas of high tritium (stippled on the map) to areas of low tritium.

10.6 The special role of tritium in tracing intermixing of old and recent waters

In sections 6.4 and 6.6 mixing of two water sources was discussed in terms of composition diagrams (Figs. 6.4, 6.11, 6.17, 6.18, and 6.19). The parameters of mixed waters plot on straight lines, but in most cases these lines are of positive correlations: for example, in the mixing of a deep warm water with a shallow cold water. In most cases the deep water also has a higher concentration of dissolved ions, and thus the deep water has higher values in all parameters. However, to obtain a negative correlation, we need a parameter that is high in one water and low, or zero, in the other water. Tritium is often such a parameter. In the example of Vals, given in Fig. 10.12, the data indicate mixing of two water types. An upper limit of 39°C is obtained by extrapolating the best-fit line to zero tritium.

Tritium serves as a sensitive check for possible intermixing of deep water with shallow water. For example, in a study of warm springs in Zimbabwe it was important to establish whether the sampled waters were indigenous to the deeper reservoirs, or intermixed with shallow water of the adjacent rivers (in few cases the spring emerges a few metres from the river bed). The tritium level in six warm springs (54-100°C) was 0.1-0.6 TU, as compared to 6-38 TU in nine river samples (Mazor and Verhagen, 1976). By comparing respective spring and river values it could be concluded that contamination by river water was less than 2%. The springs could be regarded, in this case, as single component systems.

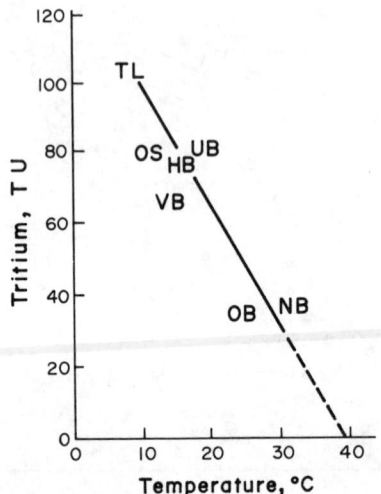

Fig. 10.12 Tritium values from various springs and wells (letters) sampled on the same day in the spa of Vals, Switzerland (Vuataz, 1982). An extrapolation of the best-fit line to zero tritium indicates an uppermost value of 39°C for the warm end member (see text).

10.7 Tritium, dissolved ions and stable isotopes as tracers for rapid discharge along fractures: the Mont Blanc tunnel case study

A unique study of rock-water interrelation has been reported from the Mont Blanc tunnel by Fontes et al. (1979). The 11.6 km long tunnel begins in a carbonate rock section in the Italian end, continues through a long section of granitic rocks, and ends in schist at the French side. The geological cross-section and chemical data are given in Fig. 10.13 and Table 10.2. A close correlation is seen in Fig. 10.13 between the lithological and chemical transects:

- High HCO_3 and pH and low Cl below the carbonate rock section, along with high Ca and Mg (samples 1-36 and 1-41 in Table 10.2).
- Low salt content in the granitic section (Fig. 10.13 and Table 10.2).
- Low HCO_3 but high and varying Cl in the schist section.

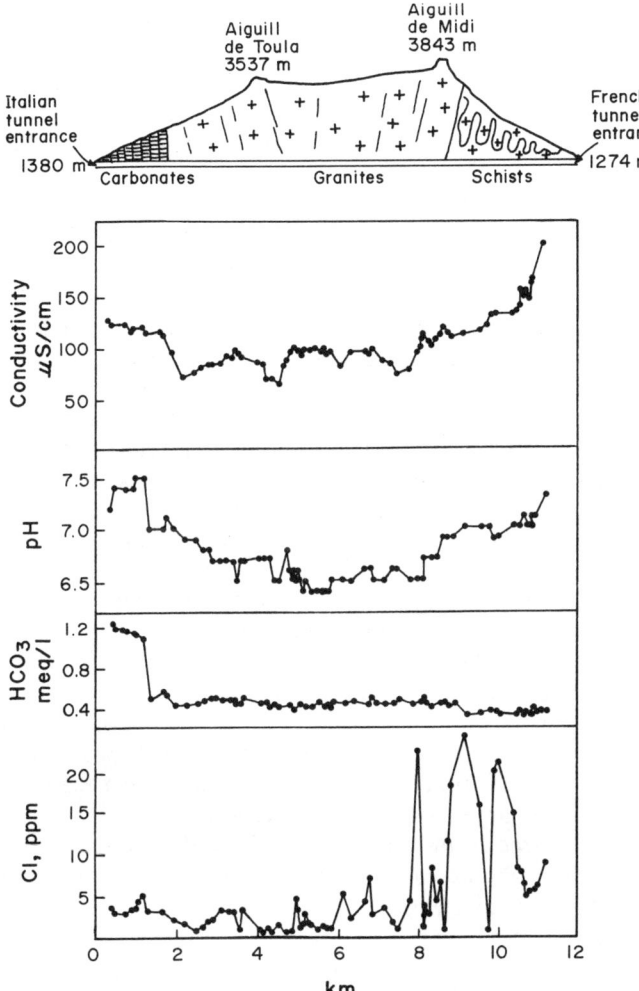

Fig. 10.13 A geological cross section and composition of waters sampled along the Mont Blanc tunnel (after Fontes et al., 1979). More data are given in Table 10.2 and in Figure 10.14. The good correlation with the lithology is discussed in the text.

Table 10.2 Composition (ppm) of representative water samples from the Mont Blanc tunnel (from Fontes et al., 1979)

Sample	Lithology	Ph	Ca^{2+}	Mg^{2+}	Na^+	K^+	Cl^-	SO_4^{2-}	HCO_3^-	SiO_2
F 1	schist	6.9	27.6	0.73	18.4	2.97	8.90	72.4	22.0	n.d.
F 27	granite	6.5	23.1	0.24	10.7	1.17	3.54	46.0	29.3	20.9
F 30	granite	6.5	9.22	0.49	13.3	1.17	4.62	26.9	25.6	14.5
F 38	granite	6.5	9.20	0.24	21.3	2.35	5.33	36.8	28.7	15.7
I 21	granite	6.7	9.20	0.49	5.04	2.74	3.55	14.1	30.5	5.6
I 36	carbonates	7.5	26.8	2.41	8.70	0.78	5.32	36.5	67.1	6.9
I 41	carbonates	7.2	32.8	2.19	9.39	2.35	3.55	31.7	75.6	n.d.

The last observation has been interpreted by the researchers as indicating the existence of evaporitic rocks in the schist complex, a conclusion that is supported also by high Ca, Na and SO_4 concentrations in the water (e.g. sample F-1 in Table 10.2).

The chemistry of water encountered in the Mont Blanc tunnel is thus directly related to the lithological section. This was interpreted as indicating *recharge along almost vertical joints, with little horizontal flow*.

The tritium and stable isotopes results are given in Fig. 10.14, along with the surface profile and geological section. The tritium data indicate rapid recharge, all tritium values being post-bomb. By comparison with the Vienna line in Fig. 10.3, the values over 200 TU may be identified as post-1961 recharge. The recharge area in the central section is around 3200 m (top of Fig. 10.14) and the tunnel is at an altitude of about 1300 m. Thus, the recharged water descended 3200 − 1300 = 1900 m in 13 years or less, giving a recharge velocity of 150 m/y, or more. The sharp variations in the tritium values indicate *recharge flow in separated fractures with varying flow velocities* (reflected by different tritium concentrations, indicating different intake dates). Comparison of the tritium data with the other parameters measured in the Mont Blanc tunnel is most instructive. The deuterium and oxygen-18 lines follow the topography (note that the δD and $\delta^{18}O$ scales were reversed in Fig. 10.14 to facilitate comparison with the topographic profile). The isotopic values are less negative at each end of the tunnel, reflecting recharge from the lower parts of the respective landscape. The central part of the tunnel hosts water with more negative isotopic values, reflecting recharge from higher altitudes.

The average recharge altitude of the ends of the Mont Blanc tunnel and of the central part may be calculated, using the isotopic altitude equations for the Maritime Alps (section 9.8):

$$h = -320 \, \delta^{18} - 2564$$

$$h = -40.2 \, \delta D - 2052$$

The δD curve in Fig. 10.14 reveals average values of −110‰ for the edges of the tunnel and −130‰ for the central part. The respective values of $\delta^{18}O$ are −15.2‰ and −17.5‰. Applying these values to the altitude equations we get (in masl):

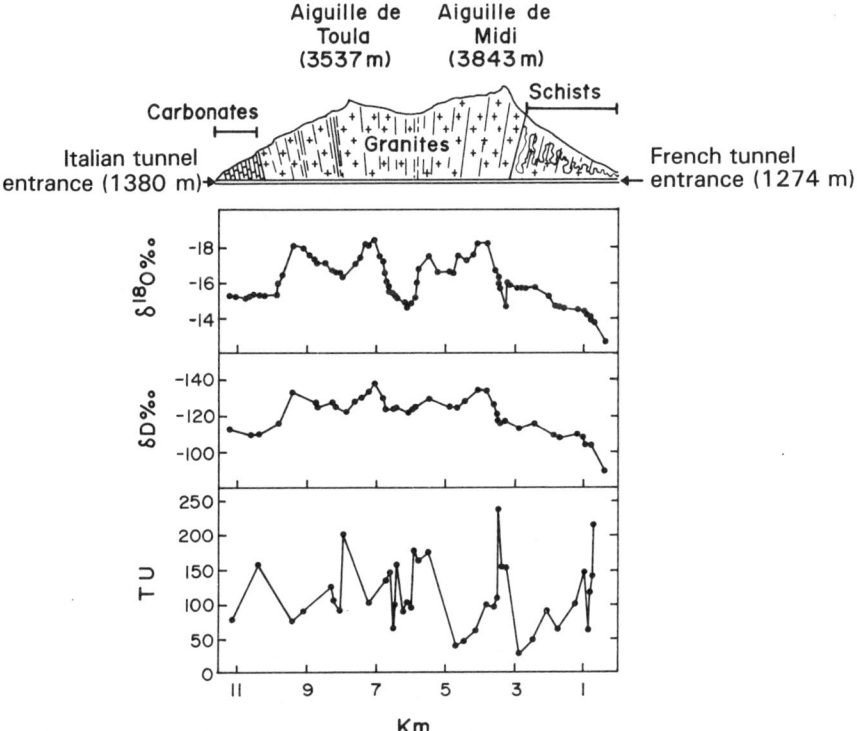

Fig. 10.14 Tritium, δD and $\delta^{18}O$ in waters collected along the Mont Blanc tunnel during its construction in 1974 (from Fontes *et al.*, 1979). The δD and $\delta^{18}O$ axes have been plotted in a reversed mode (values getting heavier downwards) to facilitate comparison with the topographic profile. The high tritium values indicate post-bomb recharge, and in the cases of over 200 TU one may even conclude post-1961 recharge (Fig. 10.3).

	hD	$h^{18}O$	mean h
tunnel ends	2300	2370	2330
central section	3170	3040	3100

These results are in good agreement with *average* altitude of the slopes of the ends, and the average altitude of the central part of the topographic profile above the tunnel (even better agreement could probably be obtained if local isotopic altitude equations were available). The stable isotopes thus indicate independently that recharged water descends in vertical fractures.

The combined application of dissolved ions, deuterium, and oxygen-18 supplied redundancy in the indications of flow in separated vertical channels and the tritium provided the time scale, or velocities.

The above study is an example of the modes by which recharge through fractures may be identified, apart from 'sponge'-type recharge. Dripping water in caves and mines is a good potential target for local studies of this type.

11 RADIOCARBON AND CARBON-13

11.1 The isotopes of carbon

Carbon has three isotopes in nature:

^{12}C - common, stable
^{13}C - rare, stable
^{14}C - very rare, radioactive, 5730 y half-life.

The heavy carbon isotope, ^{14}C, is unstable and decays radioactively into ^{14}N, emitting a beta (β^-) particle that can be measured in specialized laboratories. The half-life of ^{14}C is 5730 years. The above listed information may be summarized in the following way:

$$^{14}C \xrightarrow[T_{1/2} \ 5730 \ y]{\beta^-} {}^{14}N$$

The radioactive decay curve of ^{14}C is given in Fig. 11.1. What fraction of an initial concentration of ^{14}C is left after 10 000 years? The answer may be derived from Fig. 11.1: 27% will be left.

11.2 Natural ^{14}C production

Carbon-14 is formed in the upper parts of the atmosphere by secondary neutrons, formed by cosmic ray interactions with the atmosphere. The neutrons (n) interact with common nitrogen:

$^{14}N + n \rightarrow {}^{14}C + {}^{1}H$

The ^{14}C atoms are oxidized upon production and mix with the atmospheric CO_2. With the latter, the ^{14}C is introduced into plants and into surface water and groundwater. Most of the ^{14}C introduction into groundwater occurs through the soil CO_2 (section 2.1), which is in constant exchange with the atmospheric CO_2 and therefore has similar concentrations of ^{14}C.

The concentration of ^{14}C is expressed in relation to the $^{14}C/^{12}C$ ratio in an international standard (oxalic acid). The ^{14}C concentration in the bulk carbon of the standard is defined as 100 per cent modern carbon (pmc). Thus, a tree that died will contain only 50 pmc after 5730 y (one half-life). The ^{14}C concentration of a sample is measured in specialized low-level counting laboratories.

What is the ^{14}C concentration of a tree trunk that fell 2000 years ago? The answer, read from Fig. 11.1, is 80 pmc. The analytical limit (including contamination of the sample) of the ^{14}C measurement is in many cases about 1 pmc. What is the time range of this dating method? The answer, again read from Fig. 11.1, is up to about 40 000 y. This limit may be extended by more sensitive measuring and sampling techniques. The carbon-14 age method has been carefully studied for archeological and paleoclimatic measurements. Much effort has been invested in studying variations in the ^{14}C production over the last 100 000 years. Minor variations have been observed, but they are negligible in groundwater dating, compared to other complicating factors.

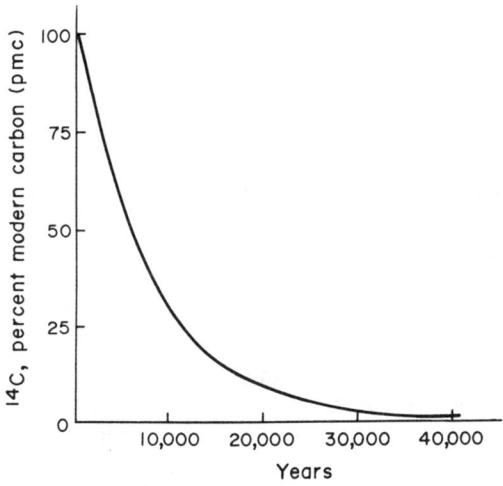

Fig. 11.1 Carbon-14 decay curve (half-life 5730 y).

11.3 Manmade carbon-14 dilution and addition

Since the industrial revolution of the early nineteenth century, large amounts of fossil fuels (oil, coal, gas) have been combusted, causing so far an increase of about 10% in the concentration of atmospheric CO_2. This added CO_2, being derived from fossil fuels, is devoid of ^{14}C and, correspondingly, has lowered the $^{14}C/^{12}C$ ratio in the air by about 10%.

An anthropogenic addition of ^{14}C into the atmosphere occurred with the nuclear bomb tests, between 1952 and 1963, along with the introduction of bomb tritium. As a result, the ^{14}C concentration also increased in plants (Fig. 11.2), in the soil CO_2 and in recharged groundwater. Values up to 200 pmc have been measured, but they decreased to about 120 pmc in 1987.

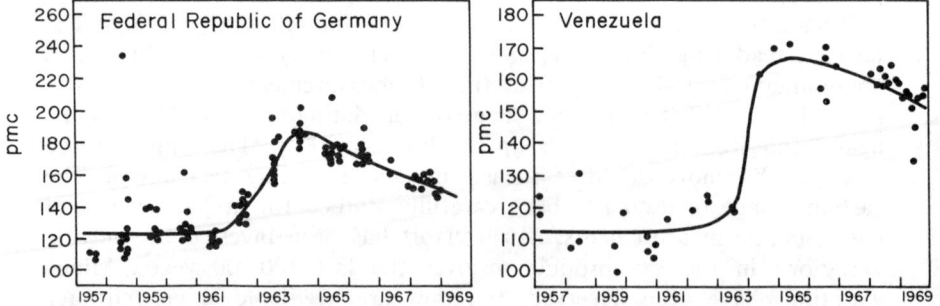

Fig. 11.2 Carbon-14 in plants in Germany and in Venezuela (Tamers and Scharpenseel, 1970). Nuclear test additions, above the natural 100 pmc, reached a peak value in 1964.

11.4 Carbon-14 in groundwater: an introduction to dating of groundwater

Rain and surface water dissolve small amounts of atmospheric CO_2. Significantly more CO_2 is added to water percolating through the soil layer, as soil air contains around 100 times more CO_2 than free air. Soil CO_2 is produced by biological action such as root respiration and decay of plant material. This CO_2 was tagged by the atmospheric ^{14}C concentrations of around 100 pmc in pre-nuclear bomb times (pre-1952), and up to 200 pmc (average: about 130 pmc) in post-bomb years.

Once water reaches the saturated zone of an aquifer it is isolated from the atmosphere and its ^{14}C decays. Table 11.1 includes ^{14}C and tritium data from wells in the Kalahari. In two cases values above 100 pmc were seen, together with post-bomb tritium (samples 2 and 14), reflecting the input without secondary changes. However, in other cases with post-bomb tritium the corresponding ^{14}C values are below 100 pmc (Fig. 11.3). This lowering is not the result of radioactive decay, as the ^{14}C half-life is too long for the decay to be noticeable in post-bomb samples. This lowering of ^{14}C is caused by interaction with carbonate rocks devoid of ^{14}C, a topic addressed in the next section.

Results of a hydrochemical study in basaltic aquifers of Hawaii (Hufen et al., 1974) are given in Table 11.2 and in a histogram in Fig. 11.4. The tritium values in the Hawaii study were 0.3–2.9 TU in 1974, indicating that most, or all, of the water had a pre-1952 age. The corresponding ^{14}C concentrations were 85–97 pmc. The pre-1952 ^{14}C in the atmosphere was a

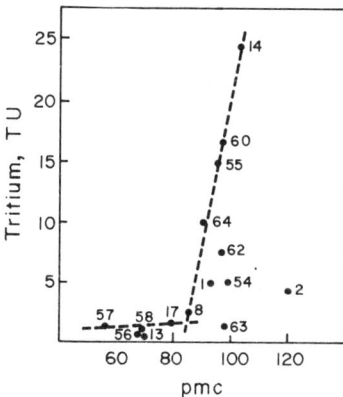

Fig. 11.3 Radiocarbon–tritium correlation, northern Kalahari groundwater (Table 11.1). Samples with values exceeding 85 pmc and 2 TU contain a post-nuclear water component. Dashed lines are theoretical dividers between pre-bomb and post-bomb waters (Verhagen et al., 1974).

Table 11.1 Carbon-14 and tritium in phreatic groundwater in the northern Kalahari (Verhagen et al., 1974)

No.	Well	^{14}C(pmc)	Tritum (TU)
2	Makgaba 2	120.2 ± 2.6	4.3 ± 0.3
14	Rakops	106.1 ± 4.5	24.4 ± 1.7
54	Khodabis 1231	99.1 ± 2.2	5.1 ± 0.6
60	Toteng 1242	98.7 ± 1.8	16.9 ± 1.7
55	Z1183	97.7 ± 2.1	15.8 ± 1.3
63	Gweta	97.6 ± 1.5	1.2 ± 0.3
62	Bushman Pits	97.4 ± 2.2	8.0 ± 1.0
1	Makgaba 1	93.7 ± 1.7	5.0 ± 0.5
64	Zoroya 1606	91.9 ± 2.2	10.3 ± 1.1
8	Steinberg's BH	86.2 ± 1.6	2.8 ± 0.4
17	Tshepe	78.4 ± 1.3	1.6 ± 1.6
11	Makoba	69.1 ± 1.3	—
13	Mahata	69.1 ± 1.3	0.7 ± 0.3
58	Tsau	68.3 ± 1.3	0.9 ± 0.2
56	Kuke	66.8 ± 1.7	0.6 ± 0.6
59	Toteng 1320	59.5 ± 1.0	—
57	Sehitwa tribal borehole	54.3 ± 1.2	1.2 ± 0.3

little lower than 100 pmc (due to the additions of fossil CO_2). Hence, part of the waters reported in Table 11.2 maintained all their initial ^{14}C as expected in non-carbonatic terrains (section 6.8). Other samples of the Hawaii study lost up to 15% of the initial value. There are two possible reasons for this loss: either interaction with carbonatic rocks devoid of carbon-14 (secondary in the basaltic terrain), or ageing of the water, reflected in radioactive decay of the ^{14}C. We will return to the Hawaii study (section 11.6), discussing additional information provided by the $\delta^{13}C$ values.

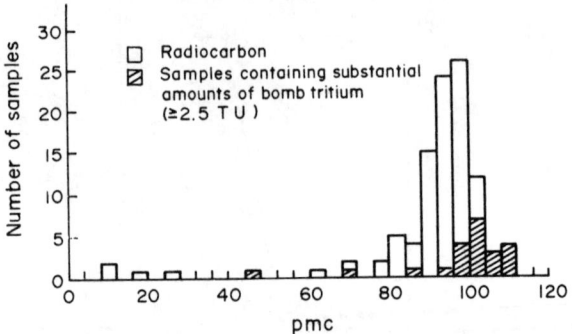

Fig. 11.4 Histogram of radiocarbon and tritium in groundwater in basaltic aquifers, Oahu island, Hawaii (Hufen et al., 1974). Pre-bomb (pre-1952) samples maintained nearly 100 pmc, indicating limited 'dilution' by ^{14}C devoid carbonates (see text).

Table 11.2 Carbon isotopes and tritium in recent groundwater in basaltic aquifers, Hawaii (from Hufen et al., 1974)

Pumping station	Draft (10^6 m^3/a)	Cl (mg/l)	HCO$_3^-$ (mg/l)	Tritum (TU)	Radiocarbon (pmc)	^{13}C (δ‰ PDB)
Aina Koa Wells	0.57	134	82	1.0	96.1	−19.2
Waialae Shaft	0.47	135	102	0.8	97.5	−17.2
Kaimuki Wells	7.43	76	76	0.6	96.1	−18.5
Wilder Wells	10.79	43	107	2.9	85.0	−15.5
Beretania Wells	12.76	52	73	0.3	91.4	−19.5
Kalihi Shaft	15.89	73	69	0.6	91.5	−18.3
Halawa Shaft	17.20	46	67	1.6	97.3	−19.2

A case study of groundwater in the Chad basin, conducted during 1967 and 1968 (Vogel, 1970), revealed a wide spectrum of ^{14}C concentrations, from 0 to 150 pmc (Fig. 11.5). Waters with over 100 pmc were clearly post-1952, demonstrating the possible use of ^{14}C as an independent short-term age indicator complementing the tritium method. The high ^{14}C values further indicate that little dilution by ^{14}C devoid carbon occurred due to water–rock interaction. Waters with very low ^{14}C levels also prevailed in the Chad study area, indicating presence of old groundwater, tens of thousands years old (several ^{14}C half-lives).

Radiocarbon and Carbon-13 169

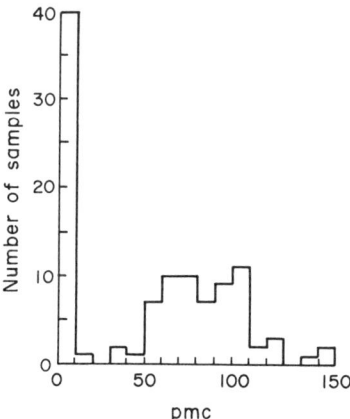

Fig. 11.5 Carbon-14 values in groundwater samples, collected during 1967 and 1968 in the Chad basin, north Africa. Post-nuclear bomb recharge is observed in several cases with above 100 pmc. Very old waters with low ^{14}C were also common (from Vogel, 1970).

11.5 Lowering of ^{14}C content by reactions with rocks

Assume that water entered the saturated zone with an initial concentration of 100 pmc, 4000 years ago, and emerges at present in a well. How much ^{14}C will it contain? The answer, read from Fig. 11.1, is 62 pmc.

What should be the ^{14}C concentration in a spring issuing in a silicate (sandstone, shale, igneous, etc.) terrain near Bonn, Germany, if the water was recharged in 1964 and emerged in 1982? From Fig. 11.2 it can be seen that due to the nuclear tests the atmospheric ^{14}C concentration rose to about 180 pmc in 1964. This concentration would not be affected to a measurable extent by radioactive decay in the 18 years that passed between recharge and emergence (the half-life being 5730 years).

The last two hypothetical examples are correct only in the case where the groundwater behaves as a closed system with regard to its $^{14}C/^{12}C$ ratio. Table 11.2 presents ^{14}C data for basalt aquifers in Hawaii, of mainly pre-bomb age, as reflected in the low tritium concentrations. The waters are seen to contain only up to 100 mg HCO_3/l and to have ^{14}C values exceeding 91 pmc in seven out of eight reported pre-bomb cases. It seems, therefore, that little or no ^{14}C was lost due to reaction with rocks and little or no 'dead' carbon (i.e. devoid of ^{14}C) was added by interactions with aquifer rocks, a fact reflected also in the low HCO_3 concentration.

In contrast, in the case of the Kalahari waters reported in Table 11.1, we noticed significant lowering of the ^{14}C content in recent, tritium-containing, waters. This is a common observation, mainly in carbonate terrains. Limestone or dolomite contain no ^{14}C because they have been formed a long time ago compared to the ^{14}C half-life of 5730 y. Recharged water containing CO_2 reacts with the carbonates to form dissolved bicarbonate:

$$H_2O + CO_2 + CaCO_2^- \longrightarrow Ca^{2+} + 2HCO_3^-$$
$$\text{100 pmc} \quad \text{0 pmc} \qquad \qquad \text{50 pmc}$$

Thus, the dissolved CO_2 with 100 pmc reacts with the rock with 0 pmc to produce a 50 pmc bicarbonate.

Similar reactions also occur with silicates:

$$H_2O + 2CO_2 + 2NaAlSi_3O_8 \rightarrow 2Na^+ + Al_2O_3 + 6SiO_2 + 2HCO_3^-$$
$$\text{100 pmc} \qquad \qquad \qquad \qquad \qquad \qquad \text{100 pmc}$$

or

$$H_2O + 2CO_2 + 2KAlSIi_3O_8 \rightarrow 2K^+ + Al_2O_3 + 6SiO_2 + 2HCO_3^-$$
$$\text{100 pmc} \qquad \qquad \qquad \qquad \qquad \qquad \text{100 pmc}$$

In such reactions, the initial ^{14}C concentration of 100 pmc (in pre-1952 samples) is closely maintained, as seen in the examples from Hawaii (Table 11.2).

A variety of reactions, of the kinds mentioned and other types, occur in the aerated and saturated zones, causing the initial ^{14}C concentration in pre-bomb groundwater to vary between 100 pmc (in silicate rocks) and 50 pmc (in carbonate rocks). In reality, the water–rock interaction in carbonate rocks causes a drop to only $60 \pm 5\%$ of the initial ^{14}C in the recharged water: values of 60 ± 5 pmc seem to be common for recent, but pre-bomb, groundwater in calcareous aquifers (Kroitoru et al., 1987). A practical way to define the initial ^{14}C that groundwater attains due to rapid water–rock interaction is to measure the ^{14}C in recent local groundwater and apply it to the age calculation of older water of aquifers with similar rocks. In the absence of such local information, a drop of 60% of the initial ^{14}C may be attributed to water–rock interaction in carbonate aquifers, and a drop to 90% of the initial ^{14}C value in silicate rocks (plutonic and volcanic rocks, sandstone, shale and quartzite).

Example. A local shallow water has been observed to contain 65 pmc and 5 TU. A sample from an adjacent deep well, in similar rocks, revealed 20 pmc and OTU. What is the deduced age of the water in the second well? The value of 65 pmc may be taken as the initial ^{14}C concentration of the groundwater when it first reached the deep aquifer. The observed 20 pmc represents $20 \times 100/65 = 30.7\%$ of the initial value, indicating an age of 10 000 years (Fig. 11.1).

The reader is referred to a number of basic papers, dealing with chemical and isotopic aspects of groundwater dating by radiocarbon: Back and Hanshaw (1970), Pearson and Hanshaw (1970), Fontes (1983), Pearson and Swarzenski (1974), Geyh (1972, 1980).

11.6 Carbon-13 abundances, and their relevance to ^{14}C dating of groundwater

So far we have discussed the occurrence of ^{14}C in hydrological systems. In a similar way one may follow the ^{13}C. Its abundance in rocks, organic material or groundwater is expressed in per mil deviation of the $^{13}C/^{12}C$ ratio in the sample from that in a standard (PDB-a Belamnite carbonate from the Pee Dee Formation of South Carolina). Most marine carbonate rocks have $\delta^{13}C = -2$ to $0‰$, whereas frequent values for organic material and CO_2 in soil are $-28‰$ to $-20‰$. Most plants have values around $-23 \pm 3‰$, but certain plants have more positive values, around $-12 \pm 2\%$ (Tables 11.3-11.5 and Figs. 11.6, 11.7).

Table 11.3 $\delta^{13}C$ values for selected samples from Saudi Arabia (Shampine et al., 1979)

Source	$\delta^{13}C$ (‰)
Tuwayq Mountain limestone	0.8
Upper Dhruma limestone	0.9
Upper Wasia Sandstone	−1.3
Carbonaceous shale, from 241 m	−25.7
Lignite, from 1450 m	−23.4
Thorn plant, living	−24.5

Table 11.4 Carbon isotopes of various samples from a forest near Heidelberg, Germany (from Vogel, 1970)

Analyses No.	Material	Collection date	$\delta^{13}C(‰)$	^{14}C (pmc)
H-415	Air CO_2	Apr 1958	−23.8	109
H−414	Soil CO_2	Apr 1958	−23.4	104
H-447	Growing leaves	Jun 1958	−29.5	109
H-655	Air CO_2	Apr 1959	−23.8	123
H-680	Air CO_2	Jun 1959	(−23)	129
H-701	Air CO_2	Jul 1959	−21.8	126
H-723	Soil CO_2	Mar 1959	(−23)	106
H-703	Growing leaves	Apr 1959	−28.1	112
H-702	Growing leaves	Jul 1959	−31.4	120
H-679	Humus 0-2 cm	Jun 1959	−25.7	94
H-749	Humus 2-5 cm	Jun 1959	−26.0	90

Table 11.5 $\delta^{13}C$ in soil CO_2 and associated plants in northern Chile (from Fritz et al., 1979)

Station	$\delta^{13}C$ (‰), soil	$\delta^{13}C$ (‰), plant
1	−14.7	−14.0
2	−22.3	−28.1
3	−18.3	−20.0 to −23.6

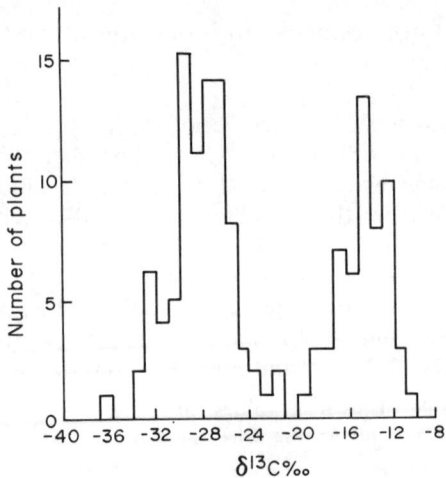

Fig. 11.6 Carbon-13 histogram in plants from Australia (after Rafter, 1974). Two major groups evolve, the total range being from $-36‰$ to $-10‰$.

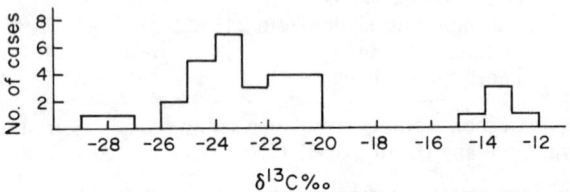

Fig. 11.7 Carbon-13 histogram in plants from Salar de Atacama, Chile (after Fritz et al., 1979). Two major populations emerge: $\delta^{13}C = -23 \pm 3‰$ and $\delta^{13}C = -14 \pm 2‰$, attributed to different systems in plant metabolism.

The concentration of $\delta^{13}C$ in groundwater is determined by the input with recharged water, and by reactions with rocks. For example, the reaction of water charged with CO_2 of $\delta‰C = -25‰$ is:

$$H_2O + CO_2 + CaCO_3 \longrightarrow Ca^{2+} + 2HCO_3$$
$$\delta^{13}C = -25‰ \quad \delta^{13}C = 0‰ \quad\quad\quad \delta^{13}C = -12.5‰$$

In contrast, in reactions with silicates the original (organic) $\delta^{13}C$ values will be retained:

$$H_2O + 2CO_2 + 2NaAlSi_3O_8 \rightarrow 2Na^+ + Al_2O_3 + 6SiO_2 + 2HCO_3^-$$
$$\delta^{13}C = -25\% \quad\quad\quad\quad\quad\quad\quad\quad\quad\quad \delta^{13}C = 125\%$$

and

$$H_2O + CO_2 + 2KAlSi_3O_8 \rightarrow 2K^+ + Al_2O_3 + 3SiO_2 + 2HCO_3^-$$
$$\delta^{13}C = -25‰ \quad\quad\quad\quad\quad\quad\quad\quad\quad\quad \delta^{13}C = -25\%$$

At this point the $\delta^{13}C$ data of the Hawaii study (section 11.4 and Table 11.2) warrant discussion. The tritium values indicated pre-1952 age, the ^{14}C values of 85-97 pmc indicated no, or little, lowering of the initial ^{14}C value ($\sim$ 100 pmc) by interaction with secondary calcite (section 11.5). The corresponding $\delta^{13}C$ values are -19 to $-15‰$. These values are closer to those of common plant material ($-23‰ \pm 3‰$) and far from those of calcareous rocks (-2 to $0‰$). Thus the ^{13}C, like the ^{14}C, changed only slightly due to water-rock interactions in the basaltic aquifers.

11.7 Application of $\delta^{13}C$ to correct observed ^{14}C values for changes caused by interactions with carbonate rocks

It has been suggested by several researchers (e.g., Pearson and Hanshaw, 1970) that $\delta^{13}C$ values may be applied to evaluate the extent by which ^{14}C in groundwater is altered by exchange with rocks. Three values have to be known:

- The $\delta^{13}C$ of the local soil material, representing the initial composition of groundwater, prior to interaction with rocks.
- $\delta^{13}C$ of the local aquifer rocks.
- $\delta^{13}C$ of the studied groundwater.

For example, if the soil has $\delta^{13}C = -24‰$, the rock $-1‰$ and the water $-19‰$ then the initial ^{14}C in the water was 'larger' (in absolute values) by a factor of:

$$\frac{-1 - (-24)}{-1 - (-19)} = 1.3$$

If the ^{14}C in the same water is observed to contain 30 pmc, then this value should be multiplied by 1.3, in order to correct for carbon-14 losses during reaction with aquifer carbonates: $30 \times 1.3 = 39$ pmc. This value is then applied to the decay curve (Fig. 11.1) and an age of 7500 years is obtained (instead of the 10 000 y indicated by the non-corrected ^{14}C value of 30 pmc). This mode of correcting ^{14}C values for losses through interaction with rocks has serious drawbacks, because the composition of ^{13}C in soil materials and aquifer rocks varies over a wide range. It is difficult, and occasionally impossible, to collect relevant samples for measurement. In some studies the more common values of $-25‰$ for $\delta^{13}C$ of soil material and $0‰$ for $\delta^{13}C$ in the rock are assumed. However, in certain cases such assumptions lead to severe overcorrections. For example, in well no. 2162 at Serowe, Botswana, the ^{14}C content was 95 pmc and $\delta^{13}C$ was $-11‰$. Hence, the correction factor would be:

$$\frac{0 - (-25)}{0 - (-11)} = 2.3$$

Applying this factor to the observed ^{14}C value in the water, a value of 95 × 2.3 = 219 pmc is obtained. This is clearly a heavy *overcorrection* (this is not a post-bomb case, as the accompanying tritium was only 0.7 TU). In this example the overcorrection is obvious as it leads to a value that is impossibly high. But if, for example, the observed ^{14}C value is 12 pmc, then such a correction would yield a value of 28 pmc, which is in the 'permitted' range and hence the overcorrection would not be noticed. As will be discussed further on, the main importance of the carbon isotopes in hydrological studies lies in:

- Providing relative ages.
- Indicating aquifer flow velocities.
- Checking the continuity of proposed regional aquifers.
- Studying mixed systems.
- Establishing modes of flow.

In these applications the $\delta^{13}C$ and $\delta^{14}C$ are most useful independent tracers.

11.8 Direction of down-gradient flow and groundwater age studied by ^{14}C: case studies

Regional aquifers consist, in certain cases, of a recharged outcrop of a conductive rock layer that dips below other rocks and becomes confined (section 2.6). Occasionally, wells tap only the confined section of such systems, but in other cases wells exist also in the suggested recharge area. In such systems carbon isotope studies are relevant to:

- Checking the suggested continuity of water flow in the aquifer.
- Checking additional recharge routes.
- Calculating flow velocities. A basic premise is that groundwater becomes older as it moves along the hydraulic gradient.

Artesian aquifer, south coast, South Africa

A case study, often quoted, has been presented by Vogel (1970) of an artesian aquifer in an area near the south coast of South Africa (Fig. 11.8). The decrease in ^{14}C down slope in the suggested aquifer has been taken by the researcher as indicating continuity. One can even calculate the velocity of groundwater flow in the aquifer, by selecting two points on the lines of Fig. 11.8, e.g. 2 km, 4000 y, and 18 km, 28 000 y. The average flow velocity in the aquifer is:

$$\frac{18 - 2}{28\,000 - 4000} \times 1000 = 0.66 \text{ m/y}$$

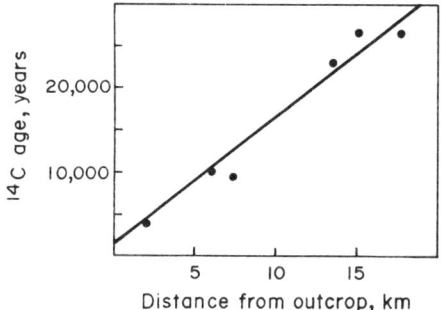

Fig. 11.8 Carbon-14 ages (assuming initial concentration of 85 pmc) of water samples from artesian wells near the south coast of South Africa, as a function of the distance from the outcrop of water-bearing strata (Vogel, 1970).

This case study is a good example to demonstrate the general idea of 'expected' behaviour and application of ^{14}C in aquifer studies. However, more data are needed in such cases:

- Closer points of sampling (in Fig. 11.8 a gap of 7 km exists between the central wells).
- Piezometric levels.
- Temperatures.
- Chemical compositions.
- Isotopic compositions, (e.g. δD, $\delta^{18}O$, $\delta^{13}C$ and tritium).

The continuity of the aquifer, and active flow in it, are independently checked by each of these parameters. This is essential, because ^{14}C may decrease with depth (and distance) also in cases of discontinuous flow. As a matter of fact, the obtained average flow velocity of 0.66 m/y is extremely slow and might indicate lack of flow in parts of the system.

Paris basin

A classical hydrochemical study of the Paris basin, which is 400 km across, has been presented by Evin and Vuillaume (1970). Their data have been compiled in Table 11.6 and the location of studied wells is plotted in Fig. 11.9. Important information included in Table 11.6 relates to the type of aquifer (phreatic, or free, versus confined), nature of the water source (spring, well, artesian flow), and the stratigraphic unit at which the respective water was tapped. A hydraulic map of piezometric level contours has been constructed (Fig. 11.10) for water in the Albian aquifer. The distinction of aquifers has always to be remembered in the processing of hydrological and hydrochemical data. Too often piezometric level contour maps are prepared of all the wells in a study area, ignoring the stratigraphic

Table 11.6 Hydrological and isotopic data of groundwaters in the Paris basin (extracted from Evin and Vuillaume, 1970)

No.	Region	Place	Aquifer	Source	Stratigraphic unit tapped	Date	HCO$_3$ meq/l	Tritum TU	δ^{13}C‰	^{14}C pmc
1	South-east	Parly-Chenons	free	well	Upp. Albian	10/66	0.3	77		94.7
2		Parly-Bernier	free	spring	Upp. Albian	10/66	3.1	27		86.7
3		Poilly	free	spring	Mid. Albian	10/66	5.1	9		91.7
4		Dracy	confined	drill	Upp. Albian	7/69	2.4	10		66.7
5		Chichery	confined	drill	Upp. & Mid. Albian	7/69	2.4	1.d.*	−13.0	58.0
		Chichery	confined	drill		4/66		1.d.	−17.7	53.2
6		Migennes	confined	drill	Upp. Albian	10/67	4.0	1.d		46.6
		Migennes	confined	drill	Upp. Albian	7/69	3.8	21		67.3
7		Neuilly	confined	artesian	Upp. Albian	3/69		1.d.	−15.0	15.9
8		Fleury	confined	artesian	Lower Albian	11/67	2.8	1.d		14.9
9		Champvallon	confined	artesian	Upp. Albian	3/69		1.d.	−12.8	13.6
10		Grand-Chaumont	free	well	Cenomanian Chalk	7/69	4.7			95.6
11		Froville	free	well	Turonian	7/69	6			72.5
12		Appoigny	confined	artesian	Barremien	3/69	3.6	1.d.	−18.3	0.6
		Appoigny	confined	artesian	Barremien	10/67	2.8	1.d.		11.6
13		Grande Paroisse	confined	artesian	Albian	7/69	2.4	1.d.		3.2
14		Montbouy	confined	artesian	Upp. Albian	10/67	5.3	1.d.		28.9
15	East	Nuisement	free	drill	Albian	6/69	4.48	77		84.3
16		Humbecourt	free	drill	Albian & Aptian	6/69		208		76.9
17		Chaudefontaine	confined	drill	Albian & Aptian	6/69	6.8	10		74.8
18		Voillecomte	free	well	Albian & Aptian	6/69	2.2	1.d.		66.3
19		Montier-en-Der	confined	drill	Albian & Aptian	6/69	5.6	37		52.8
20		Dompremy	confined	artesian	Albian	6/69	5.2	1.d.	−9.0	17.9
21		Ste Ménéhoule	confined	artesian	Albian & Jurassic	6/69	1.d.			1.4
22	South	Barlieu	confined	drill	Albian &	10/67	4.7	1.d.		82.1

#	Region	Location	Type	Source	Stratigraphy	Date				
23		Blancafort	confined	drill	Albian & Cenomanien	10/67	3.9	l.d.		42.2
24	West	Chapelle Angilon	confined	drill	Cenomanien	11/67	4.1	l.d.		34.2
25		Bernecourt	confined	drill	Rauracian	5/69	2.8		−13.2	75.5
26		Thiberville	confined	drill	Albian & Kimmeridgian	5/69	5.2		−13.0	65.3
27		Brou	confined	drill	Albian	5/69	3.6		−9.0	52.0
28		Châteadun	confined	artesian	Albian	6/69	4.9		−5.7	40.9
29	North-west	Gauciel	confined	drill	Upp. Albian	7/68	4.1		−9.8	37.5
30		Marais-Verniet	confined	well	Upp. Albian	7/68	4.4	l.d.	−14.2	36.9
31		Vernon	confined	artesian	Albian	7/68	3.2		−8.2	21.7
32		Les Logs	confined	drill	Albian & Cenomanian	8/68	8/68	4.9	−9.6	15.9
33		St. Pierre-en-Port	confined	artesian	Upp. Albian	8/68	3.3		−11.4	16.0
34		Le Trait	confined	artesian	Upp. Albian	7/68	3.4		−10.9	12.4
35		Mantes	confined	artesian	Upp. Albian	3/66	3.0		−10.9	12.4
36		Pont del'Arche	confined	artesian	Up. Albian	11/67	4.1			7.2
37		Honfleur	free	spring	Cenomanian	7/68	5.8	l.d.	−13.3	78.6
38		Incarville	confined	artesian	Jurassic to Cenomanian	5/69	7.5	12	+2.9	1.4
39		Epinay	confined	artesian	Albian	3/68	3.4		−13.2	9.7
40		Villeneuve	confined	drill	Albian	6/67	2.7	l.d.		9.1
41	Central	Achères	confined	drill	Albian	3/68	2.9		−16.3	8.9
42		Ivry	confined	drill	Albian	1/68	3.2	8.5	−13.3	7.8
43		Orsay	confined	drill	Albian	1/68	3.5		−10.6	7.3
44		Le Pecq	confined	drill	Upp. Albian	3/68	2.6		−14.3	6.4
		Pantin	confined	drill	Albian & Barremian	12/66	2.6		−15.3	
47		Issy	confined	drill	Albian	7/69	2.5	l.d.	−15.0	3.7
		Noisy le Grand	confined	artesian	Albian	3/68	3.1		−14.1	3.5
48		Noisy le Grand	confined	artesian	Albian	8/69	3.1			3.7
49		Paris ORTF	confined	drill	Albian	2/67	2.6		−12.2	3.2
		Aulnay sous Bois	confined	drill	Albian	3/68	2.9		−16.4	2.7
		Aulnay sous Bois	confined	drill	Albian	3/68	2.9	l.d.	−95	1.0
		Aulnay sous Bois	confined	drill	Albian	10/69	2.9			2.6
50		ViryChatillion	confined	artesian	Albian	8/69	3.1			2.1

* l.d.: limit of detection

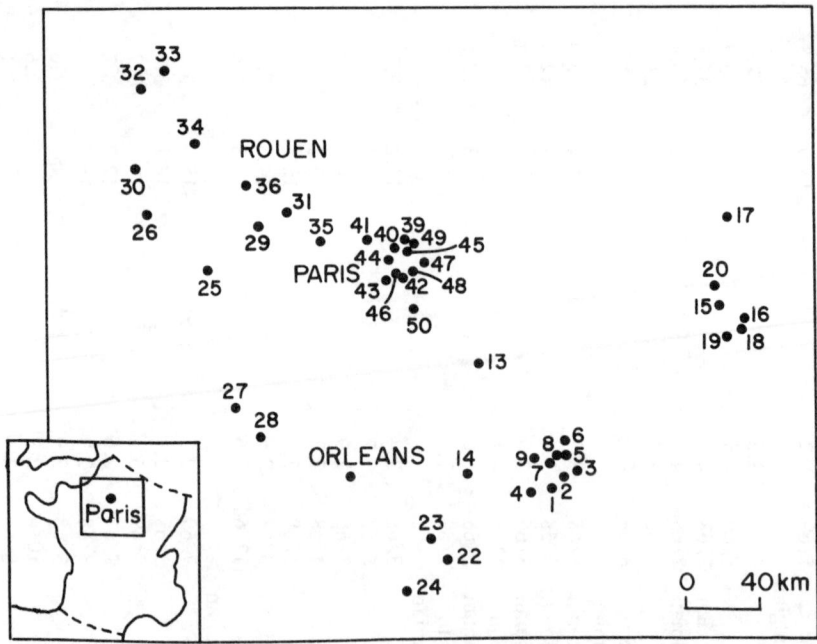

Fig. 11.9 Location map of wells studied in the Paris Basin, well numbers as in Table 11.6 (from Evin and Vuillaume, 1970).

position of the sampled water systems. In cases where water is abstracted from different stratigraphic units, water level contour maps have to be produced for each aquifer separately. The ages applied to draw Fig. 11.11 were calculated by the researchers assuming initial ^{14}C concentration of 80 pmc. The contours were drawn in ^{14}C half-life intervals. A general ageing of water from the recharge areas in the peripheries down-dip in the confined Albian aquifer is seen in Fig. 11.11, supporting the original hypothesis that was based on the piezometric data (Fig. 11.10). The Paris basin study warrants a closer look: in Fig. 11.11 it is seen that the spacing between studied wells was several kilometres near Paris, tens of kilometres at the periphery, and almost 100 km between the southern wells and the central ones. The conclusion that may be drawn is that the ageing of water towards the central part of the Albian aquifer should be regarded with extreme caution: many more wells have to be studied and local deviations may be discovered.

At this point it might be asked whether a $\delta^{13}C$-correction should be applied to the ^{14}C values in order to calculate groundwater ages in the Albian aquifer of the Paris basin. To check this point the data for the Albian aquifer (Table 11.6) have been drawn in a $\delta^{13}C$—^{14}C diagram (Fig. 11.12a). A random scatter is seen, indicating that ^{14}C dilution by interaction with carbonate rocks is not a continuous process along the down-gradient

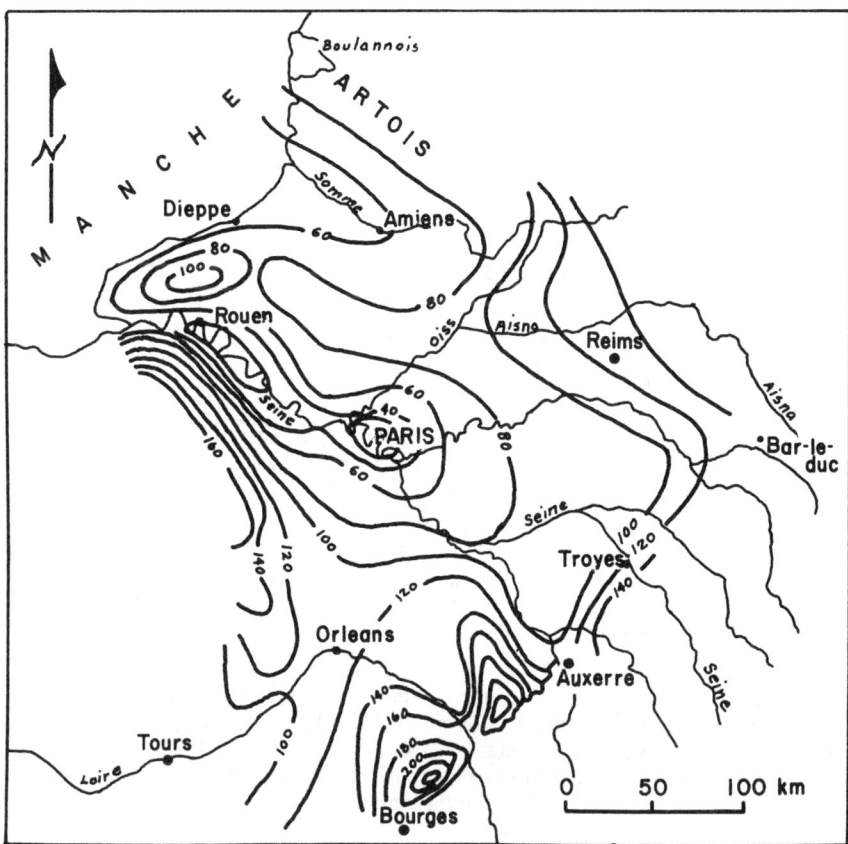

Fig. 11.10 Piezometric levels (in masl) of the Albian aquifer (after Evin and Vuillaume, 1970). Recharge seems to occur in the peripheries, giving rise to downgradient flow towards Paris in the confined Albian aquifer. This hydrological hypothesis has been tested by ^{14}C (Fig. 11.11).

flow in the aquifer. Hence, in the Paris basin no ground exists for δ^{13}C corrections.

The tritium–^{14}C values of the Albian aquifer of the Paris basin are drawn in Fig. 11.12b. Carbon-14 values in the free water table samples are in the range of 66–96 pmc, with an average of 80 pmc. This range is rather high: post-bomb ^{14}C of about 130 pmc dominated most recharge waters in the free-table aquifer, as may be deduced from the accompanying post-bomb tritium values. It therefore seems that immediate reactions with soil and rock carbonate lowered the ^{14}C concentration to a value of about 60% of the recharge concentration. This may be regarded as an initial ^{14}C concentration, applicable in ^{14}C age calculations in the confined section. A lower limit to this value may be defined by the highest ^{14}C value observed in the confined (and tritium-devoid) aquifer sample. This value is 58 pmc

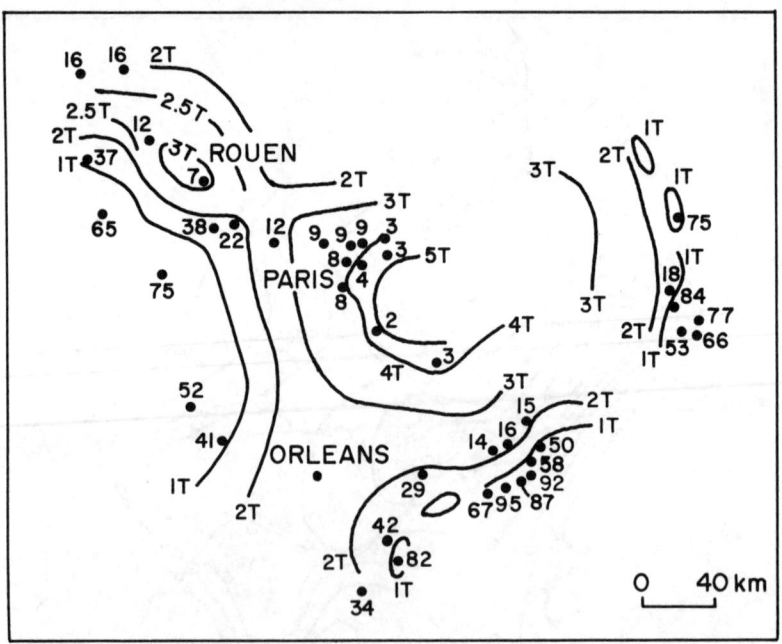

Fig. 11.11 Carbon-14 values in pmc and iso-^{14}c lines in half-life intervals (assuming an initial ^{14}C content of 80 pmc). 1T, 40 pmc; 2T, 20 pmc; 3T, 10 pmc; 4T, 5 pmc; 5T, 2.5 pmc (from data of Evin and Vuillaume, 1970). The observed increasing age towards the centre of the basin seems to confirm the hypothesis of continuous flow as postulated by the piezometric map (Fig. 11.10).

(sample 5). Thus, the initial ^{14}C value in the studied aquifer was around 60 pmc for pre-bomb recharge and 80 pmc for post-bomb recharge. The researchers in the Paris basin study selected the 80 pmc value (Fig. 11.11).

Four samples (no. 4, the July 1969 sample no. 6, no. 17, and no. 19, Table 11.6) have post-bomb tritium, yet they are from the confined part of the aquifer. Does this necessarily mean rapid flow into the confined aquifer, or may there be another explanation? These cases may represent *mixed pumping* of the confined Albian aquifer water with shallow free-table water. The Migennes (no. 6) well revealed the following sets of data:

October 1967: 0 TU and 46.6 pmc
July 1969: 21 TU and 67.3 pmc

Thus, in the 1969 pumping evidently post-bomb shallow water was added.

These examples demonstrate the high importance of tritium measurements in every ^{14}C study. In the Paris basin a relatively large number of tritium measurements was carried out and, except for the four samples named, no tritium was detected in the samples assigned to the confined aquifer.

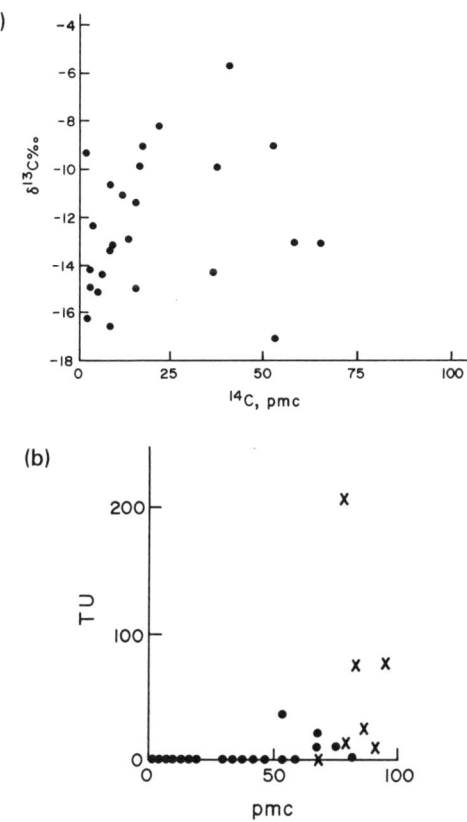

Fig. 11.12 (a) $\delta^{13}C$ as a function of ^{14}C in groundwater of the Albian aquifer, Paris basin (Table 11.6). No correlation is seen and, hence, no ^{13}C correction should be applied to ^{14}C data in the groundwater age calculations. (b) Tritium as a function of ^{14}C in the Albian aquifer of the Paris Basin (Table 11.6). ×, free water table; •, confined. Initial ^{14}C values are 58 to 80 pmc (see text).

Watrak Shedi basin, western India

Borole *et al.* (1979) presented another example of a regional aquifer with ^{14}C ages increasing down-gradient. Figure 11.13 reveals a water table map of the 500 km² basin of Watrak-Shedi, western India. The upper half of the basin was hypothesized to act as an intake area, feeding a confined aquifer existing in the downflow lower half of the basin. Carbon-14 results are given in Fig. 11.14 in terms of 'apparent radiocarbon age', i.e. an age calculated assuming an initial value of 100 pmc. Thus, waters with an initial ^{14}C value of 80 pmc will have an apparent age of 2000 y (Fig. 11.14). The upper (north-eastern) part is dominated by apparent ages of about 2000 y, representing the phreatic part of the aquifer. In the lower (south-western)

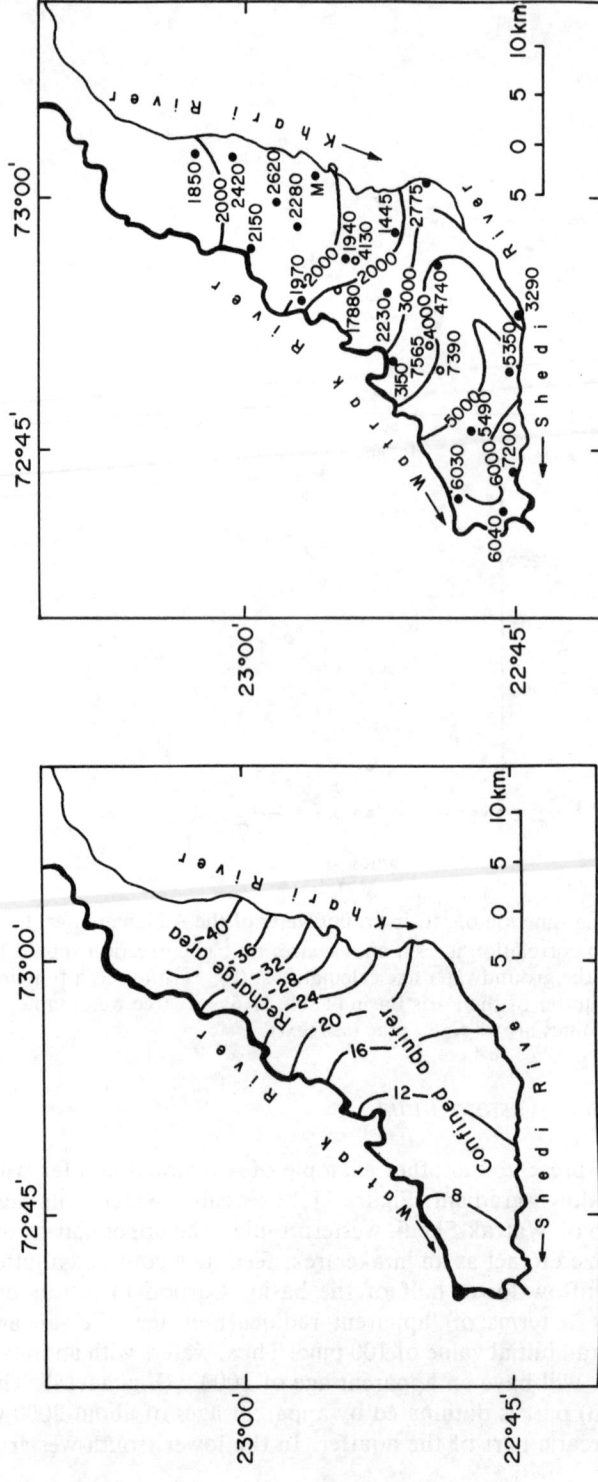

Fig. 11.13 Water-level contours of the Watrak-Shedi basin, western India (April 1978). A gradient of 0.5 to 0.6 m/km is deduced. Downgradient flow in this 500 km² basin was hypothesized (Borole et al., 1979).

Fig. 11.14 Apparent radiocarbon ages (in years, see text) in shallow (○) and deeper (●) wells, and age-contours for the Watrak-Shedi basin (Borole et al., 1979). The 2000 y values correspond to the recharge area (equivalent to initial ^{14}C of 80 pmc). The higher apparent ages indicate confinement of the aquifer in the lower half of the basin. A flow velocity of 6 m/y is calculated in the confined aquifer (see text).

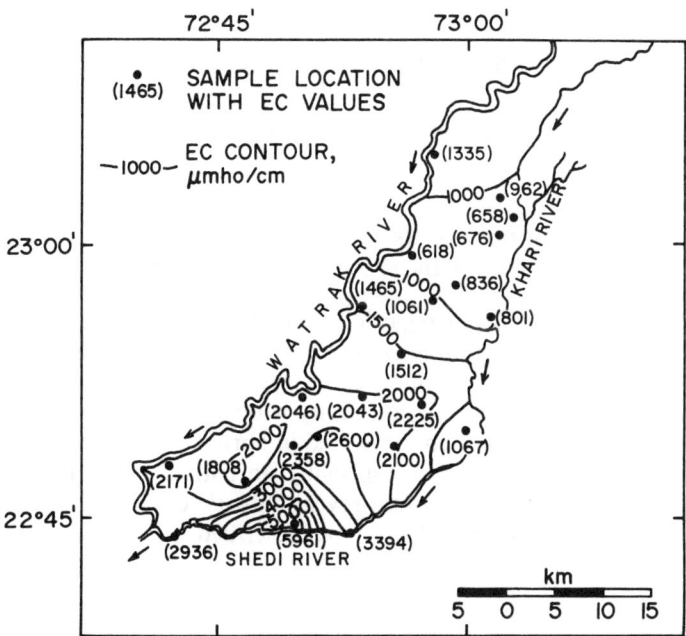

Fig. 11.15 Contours of electrical conductivity of groundwaters of the Watrack-Shedi basin (Borole et al., 1979). The phreatic section reveals fresh-water values around 1000 μmho/cm, whereas gradual salinization is observed down-gradient in the confined section.

part ages become progressively greater, interpreted as reflecting the confinement of the aquifer. What is the down-gradient flow velocity in the confined part of the aquifer shown in Fig. 11.14? It seems that the confinement begins with the lower edge of the 2000 y-apparent age zone. The distance between the lowest 2000 y contour and the 6000 y contour is about 24 km. Hence, the average velocity is:

$$\frac{24 \times 1000}{6000 - 2000} = 6 \text{ m/y}$$

The apparent age contours in Fig. 11.14 were based on data from shallow (30–80m) wells. A few values for deeper wells are given as well (circles). The age values in these deeper wells are significantly higher than in the neighboring shallow wells. This was interpreted by the researchers as indicating little or no downward mixing of the subaquifers. A good agreement is therefore observed between the hydrological model, based on the water table data (Fig. 11.13), and the observed ^{14}C data (Fig. 11.14). The down-gradient flow in the confined aquifer was further tested by electrical conductivity data (Fig. 11.15). The values indicate fresh water (1000 μmho/cm) in the phreatic aquifer, and gradual salinization in the confined section. This might well indicate an increase in salinization due to longer times of contact (and less flushing) of aquifer rocks.

The cited examples of down-gradient flow in aquifers are characterized by gradual variations in the water composition, as demonstrated by ^{14}C. These cases reveal a piston flow regime, in which flow through high-conducting zones is negligible, as otherwise the progressive change would not be maintained.

11.9 Flow discontinuities between adjacent phreatic and confined aquifers, indicated by ^{14}C and other parameters

Phreatic aquifers are often regarded as recharge zones feeding adjacent confined systems. A continuous through flow is commonly envisaged, controlled by (and deduced from) water level gradients and transmissivities. However, in certain cases a discontinuity is observed between the phreatic and confined parts of a system, reflected in abrupt changes in the chemical and isotopic compositions and, especially, in tritium, ^{14}C, ^{4}He and other age indicators. Six case studies that reveal phreatic-confined discontinuities are briefly discussed below. They were described in detail in Mazor and Kroitoru (1987).

The Judean Mountains, central Israel (Fig. 11.16)

This is an anticlinal structure of marine carbonates of the Judean Group (Cenomanian-Turonian), covered on the flanks by chalk and marl of the Mount Scopus Group (Senonian) (Fig. 11.17). Average annual rain on the mountain crest is 550 mm/y, on the Hashphela foothills to the west the value is 450 mm/y and in the Judean Desert, on the eastern slope, the value drops to 200 mm/y. Limestone, dolomite, and marl beds of the Judean Group are exposed on the mountain crest, acting as a main recharge area and hosting a phreatic groundwater system, penetrated by many wells. The Judean Group rocks are covered on the anticlinal flanks by chalk and marl of the Mount Scopus Group. The latter formation hosts limited phreatic water systems of its own, but confines the Judean Group water system. A recent study (Mazor and Kroitoru, 1987) revealed post-bomb tritium and carbon-14 values in the phreatic Judean Group system, but no tritium and low carbon-14 in the confined Judean Group system. The transitions are abrupt, both on the western and eastern flanks, as seen in Fig. 11.17.

Constancy in the δD and $\delta^{18}O$ data (Fig. 11.17) indicates that the phreatic and confined waters both originate from the same mountain crest rainy recharge zone. Thus, the phreatic and confined zones seem to communicate hydraulically, but the drainage rate of the confined sections seems to be restricted, as indicated by high ^{14}C ages.

The validity of ^{14}C as an age indicator in the described study has been established by the observed constancy of $\delta^{13}C$ values (12 ± 1‰, Fig. 11.17) and of HCO_3 and other components. The confined water has calculated ages of several thousand years (Kroitoru et al., 1987).

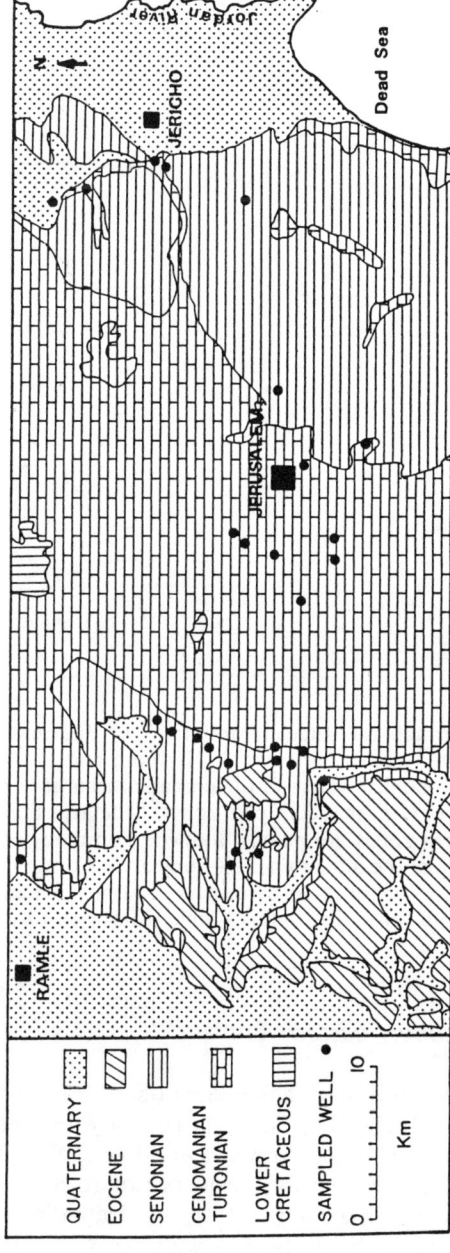

Fig. 11.16 Study area of the Judean Mountains, central Israel (Mazor and Kroitoru, 1987). Limestone and dolomite (Cenomanian-Turonian) outcrops serve as recharge areas into a phreatic aquifer, confined on the flanks by younger chalk (Senonian).

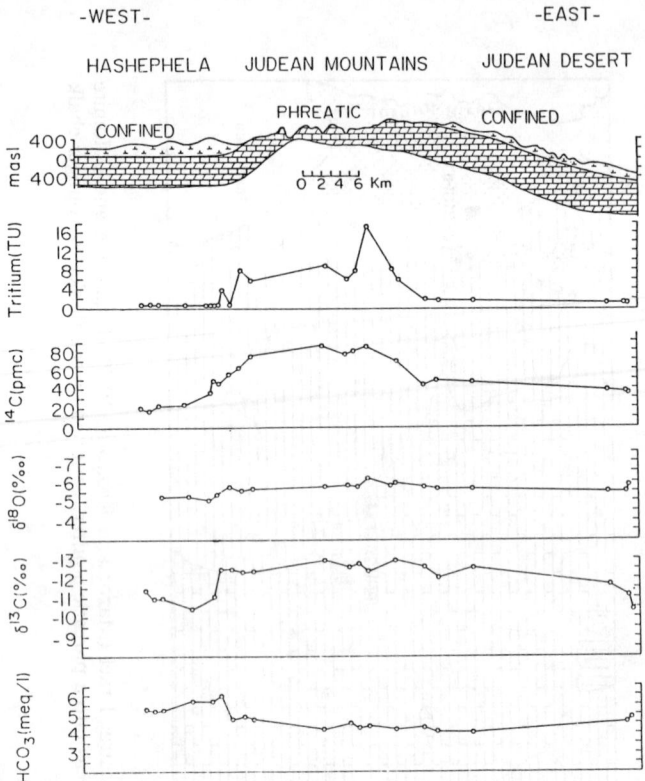

Fig. 11.17 Transects through the Judean Mountains of Judean Group rocks, partially covered by Mount Scopus Group rocks. Tritium and ^{14}C values in well samples reveal recent (post-bomb) waters in the phreatic recharge zone, and water several thousands years old in the confined zones. The transition is abrupt, indicating that the confined zones are poorly drained. Constancy of $\delta^{18}O$ values indicates common recharge at the mountain crest rainy area. Constancy of the $\delta^{13}C$ values (12 ± 1‰) indicates absence of secondary isotopic exchange reactions, proving ^{14}C is a reliable tool for dating in this system (Mazor and Kroitoru, 1987).

Bunter sandstone aquifer, eastern England

The Bunter sandstone water system has been studied over an area of 2000 km² in eastern England (Fig. 11.18). A large body of data has been published by Bath *et al.* (1979) and Andrews and Lee (1979). The regional geology, shown in Fig. 11.19, reveals a sequence of formations from the Permian in the east to the Jurassic in the west, the strata dipping eastward. The more detailed geology of the study area, with sampled well locations, is given in Fig. 11.20 and a geological cross-section is given in Fig. 11.21. The researchers suggested that water is recharged at the Bunter sandstone

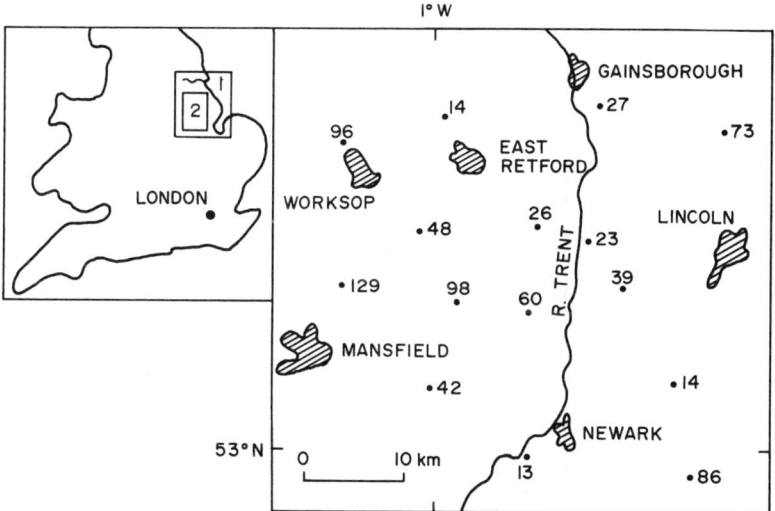

Fig. 11.18 Study area of the Bunter sandstone in eastern England, with altitude of selected points in metres. Insert: 1, area shown in Fig. 11.19; 2, area shown in Fig. 11.20.

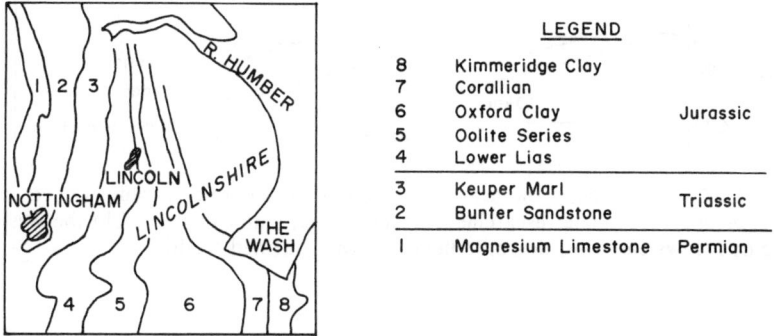

Fig. 11.19 Generalized geological map of a section of eastern England.

outcrops and moves eastwards, down-gradient, into the confined section.

The reported data have been plotted in composition diagrams in order to check whether we are really dealing with one continuous water aquifer. The Cl-δ^{18}O plot (Fig. 11.22) reveals three data clusters, indicating that three specific water groups occur. A, B, and C. The Cl-^{14}C plot (Fig. 11.23) also reveals three groups, but the data of group A are spread along a line, interpreted as a mixing line of a recent saline component (possibly contaminated with anthropogenic Cl) and a chloride-poor end member, several thousands years old. The three geochemical groups have a simple

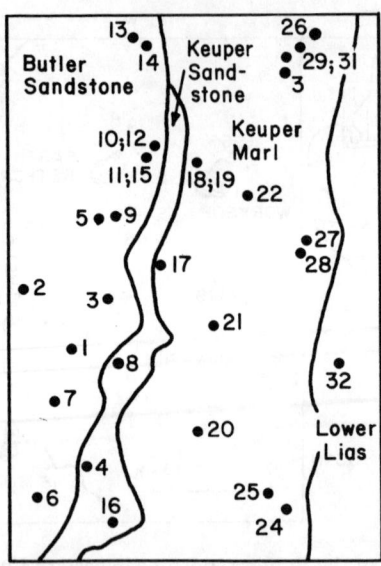

Fig. 11.20 Geology and location of sampled wells in the study area of Bunter sandstone aquifer (after Bath et al., 1979).

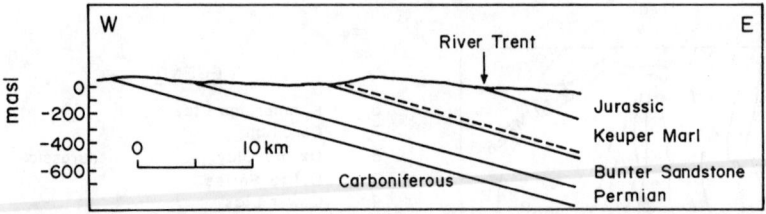

Fig. 11.21 Suggested geological cross-section of the Bunter sandstone aquifer (Bath et al., 1979). The researchers suggested that water recharged in the western outcrops flows eastward downgradient into the confined section.

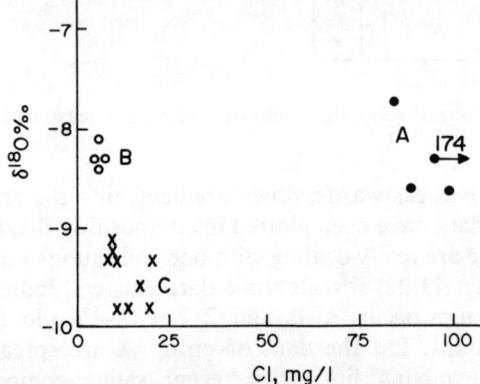

Fig. 11.22 Cl-δ^{18}O plot of the Bunter sandstone wells. Three distinct geochemical groups emerge: A, B and C (data from Bath et al., 1979).

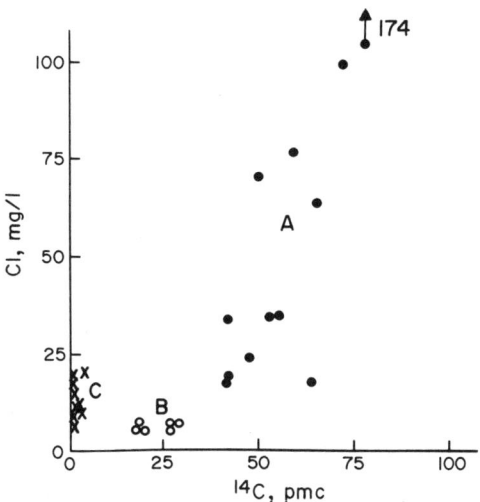

Fig. 11.23 Cl-^{14}C plot of the Bunter sandstone wells. Again, three groups emerge. The group A points plot along a line, interpreted as a mixing line of recent (post-bomb, also tritium-containing) saline (polluted?) water with several thousand year old fresh water (data from Bath et al., 1979).

geographical distribution pattern of north-south strips, as seen in Fig. 11.24. Group A coincides with the belt of phreatic Bunter sandstone outcrops, whereas groups B and C are in the confined, eastwards dipping, section. The ^{14}C, Cl and He data maps of Fig. 11.24 reveal two discontinuity lines, marked I and II. These lines are parallel with the strike of the rock strata and formations contacts (Figs. 11.20, 11.21). Thus, there is little or no active through flow from the phreatic zone of group A to the confined one of group B. Furthermore, a discontinuity also occurs in the confined zone, and the group B section does not communicate freely with the C section.

The Bunter sandstone δD and δ^{18}O data provide further support for the lack of communication between the phreatic and confined aquifer sections. The data plot along a line in Fig. 11.25, but reveal an order of B→A→C. In other words, we are actually dealing with three distinct water groups, placed on the meteoric line.

The three water groups of the Bunter sandstone have distinct ages, as revealed by two independent age indicators, ^{14}C and ^{4}He (Fig. 11.26). The positive ^{4}He-^{14}C correlation gives the age grouping a high degree of confidence. From Fig. 11.26 it seems that water group C has been separated from group B for a long time, but the discontinuity between A and B is also maintained in Fig. 11.26.

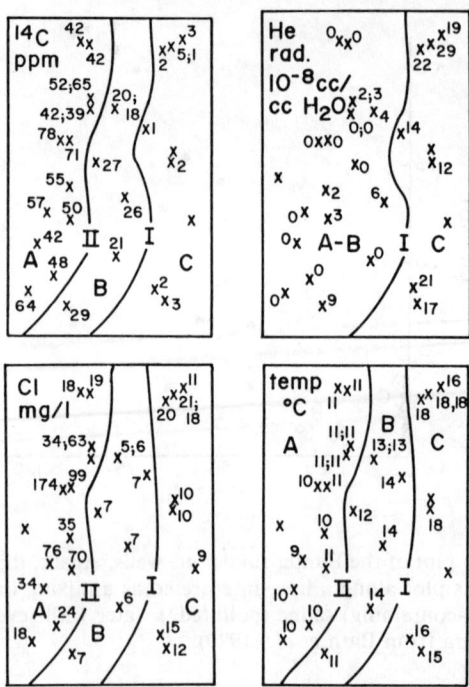

Fig. 11.24 Carbon-14, helium, chloride, and temperature data marked on the well location map of Fig. 11.20. The three distinct geochemical groups A, B, C, already defined in Figs. 11.22 and 11.23, have a clear geographical pattern: Group A coincides with the phreatic wells in the Bunter sandstone outcrop area, whereas groups B and C are in the confined section. The data reveal two discontinuity lines, I and II, discussed in the text (data from Bath et al., 1979).

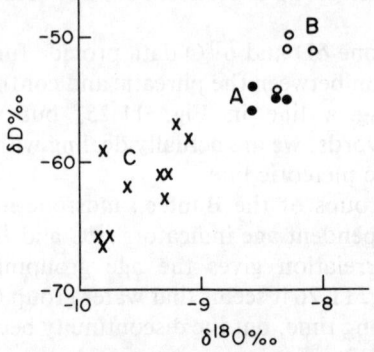

Fig. 11.25 δD–$\delta^{18}O$ plot of the Bunter sandstone wells. The pattern obtained is discussed in the text (data from Bath et al., 1979).

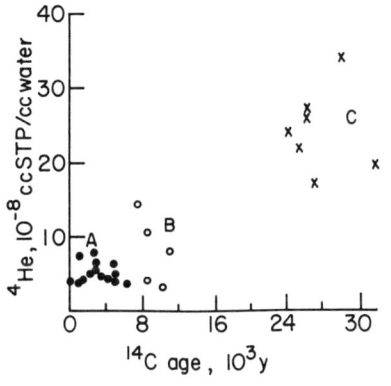

Fig. 11.26 ^{4}He–^{14}C ages of waters in the Bunter sandstone wells. Carbon-14 ages were calculated from data of Bath et al. (1979), applying 70 pmc as the initial ^{14}C concentration. The positive correlation validates the relative dating by the two independent methods. Waters of groups A, B, and C have been recharged at different times but do not represent continuous downgradient flow (see text).

Lincolnshire limestone Jurassic aquifer, eastern England

Isotopic data were reported (Dowing et al., 1977) from a number of wells in this system, located in the intake area, and in part of the confined westward-inclined section. A distinct discontinuity (dashed line) is observed by ^{14}C (Fig. 11.27). This is supported by four other parameters: tritium, δ^{13}C, δD, and δ^{18}O, reported by the researchers. Thus, little or no hydraulic communication takes place between the phreatic and confined sections of this Lincolnshire limestone aquifer.

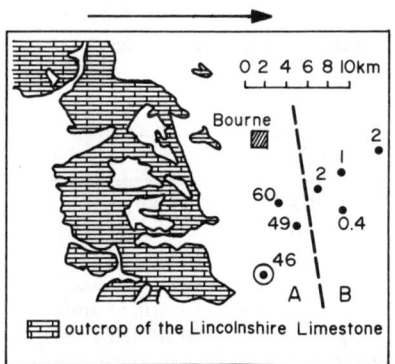

Fig. 11.27 Set of ^{14}C data (pmc) marked on well locations in the Lincolnshire limestone aquifer (data from Dowing et al., 1977). Dashed line marks discontinuity between phreatic (A) and confined (B) aquifers.

Gwandu Formation, Sokoto Basin, northern Nigeria

This system has been studied by Geyh and Wirth (1980). Figure 11.28 depicts the outcrop areas. The ^{14}C data reveal a sharp discontinuity of waters with 32–70 pmc near the outcrops (A), and 2–7 pmc in the suggested downflow section of the confined aquifer (B). This is another example of lack of hydraulic communication between adjacent phreatic and confined systems.

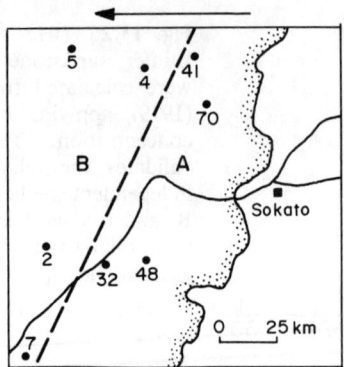

Fig. 11.28 Carbon-14 values (pmc) in wells of the Sokoto Basin, northern Nigeria (data from Geyh and Wirth, 1980). The phreatic aquifer (A) coincides with the rock outcrops (stippled) and has high ^{14}C values. An abrupt drop of ^{14}C values is observed in the adjacent confined section of the system (B).

The Florida limestone aquifer

This provides another example of phreatic–confined interrelationship. Data are given in Table 11.7. A clear phreatic–confined discontinuity is established by ^{14}C, $\delta^{13}C$ and tritium data.

Table 11.7 Isotope data for wells in the Florida limestone aquifer (Pearson and Hanshaw, 1970)

Well	^{14}C (pmc)	$\delta^{13}C$ (‰)	Tritium (TU)	Remarks
Weeki Vachee	62.4	−13.7	103 ± 10	Recharge zone
Lecanto #5	50.0	−11.5	36 ± 4	Recharge zone
Lecanto #6	51.1	−11.8	15 ± 1	Recharge zone
Frost proof	6.7	−9.8	0 ± 1	Confined zone
Holopane	5.2	−9.6	1 ± 1	Confined zone
Arcadia	3.0	−8.3	0 ± 1	Confined zone

The Blumau aquifer, southeastern Austria

This has been intensively studied (Andrews *et al.*, 1984). Here, too, drastic discontinuities were reported for groundwater properties passing from the phreatic part of the aquifer to the confined part. The transition is observed between two wells, about 400 m apart. For example, ^{14}C drops from 114 pmc to 48 pmc, tritium drops from 49 TU to 0.8 TU, $\delta^{18}O$ drops from −8.6‰ to −9.3‰, and bicarbonate rises from 100 mg/l to 320 mg/l.

Conclusions

These examples of phreatic-confined discontinuities shed light on a number of basic hydrological features:

- Flow velocity of water in an aquifer, calculated by gradients and transmissivities, provides a maximum possible value. This is subject to limitations imposed by lack of drainage. In extreme cases, confined systems may be rich in fossil karstic conduits, but with no through flow, due to complete sealing from a former base of drainage.
- Restricted drainage may, in certain cases, explain discrepancies between hydrologically calculated ages, using gradients and transmissivities, and much higher ages obtained from isotopic age indicators, e.g. ^{14}C, ^{4}He or ^{36}Cl. In fact, such discrepancies might be used as indicators for poor drainage of confined systems.
- Phreatic systems that do not drain into adjacent confined sections must have their own outputs in order to maintain the observed young ages. These outputs have to be identified to gain a full understanding of regional hydrology.
- Of special interest is the degree of abruptness of the discontinuities. These may be enhanced by clogging of rock pores in the transition zone. In other cases, a geological barrier may be present, coinciding with a suggested phreatic-confined boundary.

11.10 Mixing of groundwaters, revealed by joint interpretation of tritium and ^{14}C data

Mixing of groundwater was discussed in sections 6.6 and 6.7. There, mixing patterns were recognized by processing data from numerous samples (Figs. 6.6, 6.9, 6.17, 6.18). Mixing may also be recognized by data obtained in a single sample – in cases of 'forbidden' combinations of tritium and ^{14}C data.

Example. 1.2 TU and 48.3 pmc were found in well 1572, Hanahai, Botswana. What is the age of this water? The values represent a 'forbidden' combination, as 1.2 TU indicates a post-bomb age of a few decades, whereas 48.3 pmc indicates water a few thousand years old. In other words, the tritium and ^{14}C ages are discordant (out of agreement) in this case, and neither of the two is correct. Such a combination could be formed only by intermixing of old water (zero tritium and $<$ 48.3 pmc ^{14}C) with recent water (post-bomb tritium and post-bomb ^{14}C). The last example is of outmost importance. *Concordant ages* (ages that agree) confirm each other, and *discordant ages* indicate mixing of water of different ages.

In the identification of intermixing water types, and the calculation of mixing percentages, ^{14}C is useful because it is often the only parameter that occurs in a linear negative correlation with other parameters, needed to extrapolate end member properties (section 6.7). Deep groundwaters often have high parametric values: high dissolved ion concentrations, and elevated temperatures (as compared to the intermixing shallow and cold end member). However, ^{14}C is commonly low in the deep end member (which is old) and is high in the shallow end member (which is young), thus providing the required negative correlations. Tritium can be used in the same way (section 10.6) but its use is limited to mixtures of old water with post-bomb water. If the young end member is of a pre-1952 age it contains no measurable tritium but high concentrations of ^{14}C making the latter useful in studying mixed waters.

Table 11.8 Dissolved ions (mg/l) in springs A to D, Hammat Gader, and extrapolated fresh (fem) and saline (sem) endmembers (Mazor et al., 1973)

Spring	Li	Sr	K	Na	Mg	Ca	Br	Cl	SO$_4$	HCO$_3$
sem	0.23	17	33	375	48	196	8.8	800	262	272
A	0.143	10.4	21.8	245	39.8	152	5.4	506	177	330
B	0.079	6.0	13.1	140	44.5	105	3.7	300	121	344
C	0.057	4.1	10.5	120	37.5	124	—	218	100	361
D	0.010	1.3	3.9	55	32.7	89	2	87	58	381
fem	0.0	0.0	0.9	18	34	78	0.0	10	37	385

An example of the use of ^{14}C in studying a mixed groundwater system is provided by the Hammat Gader spring complex, in the Jordan Rift Valley, south-east of the Tiberias Sea (Table 11.8). Four springs emerge at Hammat Gader, along a stretch of a few hundred metres. Various parameters have been plotted on compositional diagrams, as a function of Cl concentration (Fig. 11.29). Linear positive correlations are exhibited by Li, K, Na, Mg, Ca, Sr, SO$_4$ and Br. These positive linear correlations indicate mixing of a fresh water with a more saline water. Positive correlations with Cl were observed for three additional parameters: temperature, dissolved He, and radium (Fig. 11.30). A negative correlation was provided in the Hammat Gader study by ^{14}C (Fig. 11.31). The vertical axis in Fig. 11.31 deserves some discussion: the ^{14}C data, reported in pmc, provide the value of the ^{14}C/^{12}C ratio in the sample, compared to a standard (section 11.2). In hydrochemical jargon the pmc values are often called ^{14}C 'concentrations'. However, to obtain values that are really concentration-related, the pmc value has to be multiplied by the corresponding HCO$_3$ concentration, as shown in Fig. 11.31. The best-fit line is seen in Fig. 11.31 to extrapolate to 69°C, providing the maximum possible temperature of the warm end member. The real temperature of the warm end member lies in the range between the highest temperature measured (52°C) and the extrapolated value (69°C). The extrapolated temperature of the cold end member was obtained from Fig. 11.30, assuming a Cl concentration as low as that of

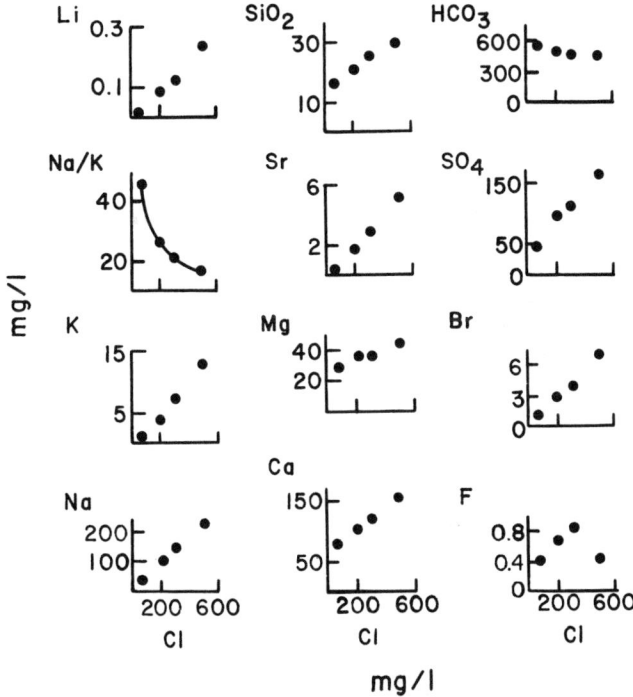

Fig. 11.29 Composition diagrams of four springs issuing a few hundred metres apart, Hammat Gader, Jordan Rift Valley, Israel (Mazor et al., 1980). Mixing of two water types is evident.

local rain water, 10 mg/l. An extrapolated temperature of 25°C has been obtained for the cold end member. The real value lies between this extrapolated value of 25°C and the lowest measured value, 29°C. The extrapolated Cl values could then be applied to the positive correlation lines of Fig. 11.29 to calculate the concentration of the dissolved ions in the two end members.

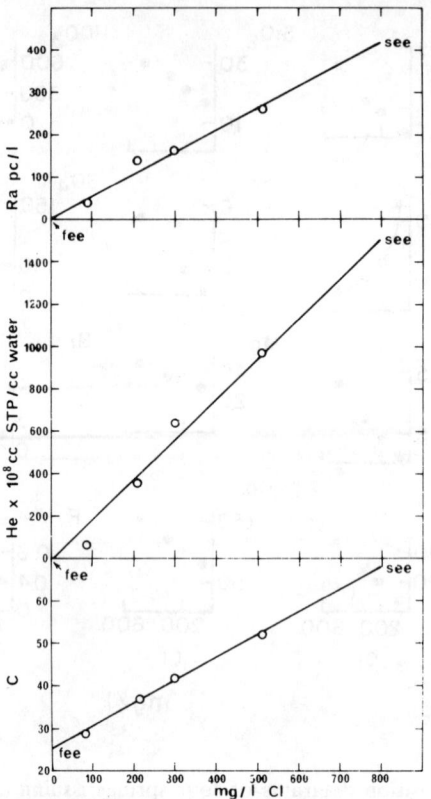

Fig. 11.30 Temperature, helium and radium as a function of chloride, Hammat Gader springs A to D and extrapolated end members (see text) (Mazor et al., 1973).

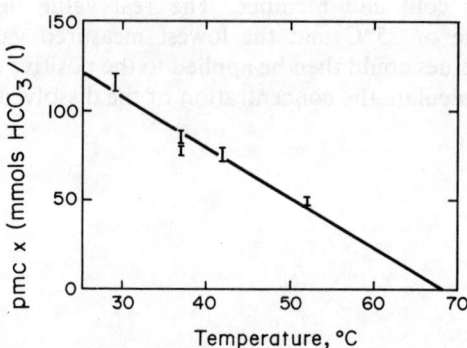

Fig. 11.31 Carbon-14 versus temperature in the spring complex of Hammat Gader. The negative correlation could be used to extrapolate to 0 pmc, obtaining a value of 69°C. The real hot end member lies between the highest value measured (52°C) and the extrapolated value (69°C) (Mazor et al., 1973).

11.11 Piston flow versus karstic flow, revealed by ^{14}C data

Piston flow is indicated by a 'layered' occurrence in the porous rocks of the saturated zone of the intake area: younger water layers overlie older water layers. Such systems are indicated by downwards decreasing ^{14}C values in depth profiles. In contrast, in karstic systems of water descending in conduits crossing porous rocks, occasionally 'disordered' depth profiles occur, with water of relatively high ^{14}C values underlying water with lower ^{14}C values.

Case studies near Nyirseg, eastern Hungary, provide examples for the discussed types of depth profiles (Table 11.9). The first two wells, of Fehergyarmat and Nyirlugos, reveal 'normal' depth profiles, with ^{14}C decreasing downwards, indicating water layering, which in turn, indicates a piston flow of recharge water. The following five wells reveal disordered depth profiles, e.g. Berettyoujfalu which in a sequence of depth profile samples gave the following ^{14}C values: 85.1, 1.2 and 15.6 pmc. These five wells encountered karstic conduits, or zones of preferred conductivity, with waters of varying ^{14}C values.

For a review of other methods that have been developed to date groundwaters, the reader is referred to a paper by Davis and Bentley (1982).

Table 11.9 Carbon-14 depth profiles in wells near Nyirseg, eastern Hungary (from Deák, 1979)

Location	Perforation (m under surface)	$\delta^{13}C$ (‰) PDB	Tritium (TU)	$^{14}C \pm \sigma$ (pmc)
Fehérgyarmat	÷10	−15.3		78.7 ± 6.3
	41− 56	−16.1	<6	12.8 ± 1.2
	81− 96	−17.4		12.1 ± 0.9
Nyirlugos	<10	−12.9	150	57.9 ± 1.0
	31− 34	−13.8	<6	27.6 ± 1.1
	110−175	−15.1		24.9 ± 1.1
Nyirbátor	<10	−17.2	119	91.2 ± 4.2
	46− 73	− 3.1	<6	54.7 ± 3.1
	161−193	−13.7		14.0 ± 0.8
	229−268	−13.2		16.3 ± 0.6
	297−313	−13.0		22.6 ± 0.5
Nyiradony	<10	−13.5	240	72.3 ± 3.1
	55−61	−15.3	<6	23.6 ± 2.3
	127−133	−13.1		16.4 ± 0.9
	172−219	−13.2		1.8 ± 1.8
Debrecen	70− 93	−14.5		3.4 ± 1.3
	138−181	−15.6		8.0 ± 0.5
	160−190	−12.7		12.6 ± 1.2
Derecske	154−159	−14.9		4.5 ± 0.9
	233−253	−13.5		7.9 ± 1.4
Berettyóujfalu	<10	−15.2	189	85.1 ± 2.4
	131−137	−22.6		1.2 ± 0.5
	296−355	−20.7		15.6 ± 2.2

12 NOBLE GASES

12.1 Rare, inert, or noble?

Helium, neon, argon, krypton and xenon occupy a special place in the periodic table – their outer shell of electrons is complete (2, 8, or 18) and therefore they do not interact with other atoms (section 5.4). This property brought these elements the titles of *inert* or *noble*. They are gases, present in small quantities in air (Table 12.1), and hence they have been called *rare* gases. In recent years it has been pointed out that argon is present in air in a concentration of nearly 1% and therefore the term 'rare' seemed inappropriate. A few xenon and krypton compounds have been prepared under specific laboratory conditions, and the term 'inert' has also been challenged. In nature, however, their chemical nobility is well preserved, making them ideal tracers in hydrological systems for a number of reasons:

- They are not involved in any chemical or biological activity.
- They enter groundwater from two distinct major sources: by equilibration with air during infiltration, or from deep-seated origins, which include flushing of radiogenic products out of aquifer rocks, and mantle-derived gases. The origin from these different sources is identifiable by the isotopic composition.
- The initial concentrations of atmospheric noble gases in recharge water may be calculated from the ambient annual temperature and altitude of the suggested recharge area.
- The fact that there are five noble gases provides valuable redundancy.

Let us begin the discussion with the atmospheric input.

Table 12.1 Noble gases (ppm volume) in dry air at 1 atmosphere (sea level)

He	Ne	Ar	Kr	Xe
5.24	18.18	9340	1.14	0.086

12.2 Atmospheric noble gases

The atmosphere is a well-mixed reservoir with known concentrations of noble gases (Table 12.1). These atmospheric noble gases have characteristic isotopic abundances that are given in Table 12.2. The solubility of the noble gases is given in Fig. 12.1, expressed in cc STP noble gas/cc water. STP stands for standard temperature (0°C) and pressure (760 mm Hg = 1 atmosphere).

Table 12.2 Isotopic abundances of atmospheric noble gases (chart of the nuclides, 1977)

Element	He		Ne			Ar		
Isotope	3	4	20	21	22	37	38	40
(%)	0.0014	100	90.51	0.277	9.22	0.337	0.063	99.60

Element	Kr					Xe					
Isotope	80	82	83	84	86	128	129	130	131	132	134 136
(%)	2.25	11.6	11.5	57.0	17.3	1.91	26.4	4.1	21.2	26.9	10.4 8.9

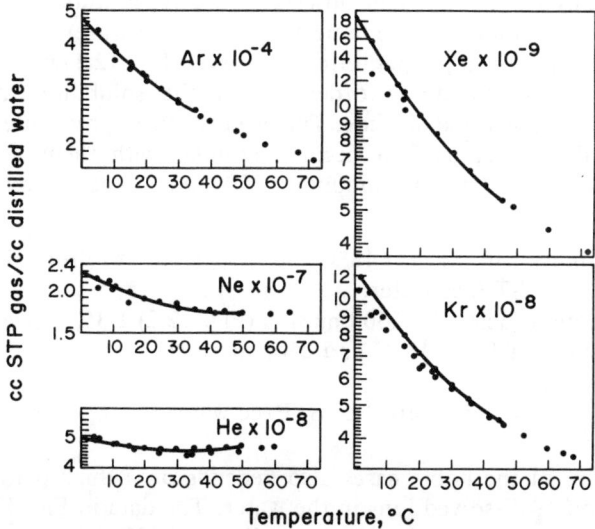

Fig. 12.1 Solubility of atmospheric noble gases in fresh water at sea level (1 atm), as a function of the ambient temperature (from Mazor, 1979).

What is the solubility of argon in distilled water, at sea level, at 15°C? The answer, from Fig. 12.1, is 3.5×10^{-4} cc STP argon/cc water.

The lack of chemical interaction of noble gases results in their behaviour as ideal gases, and as such their solubilities in water are directly

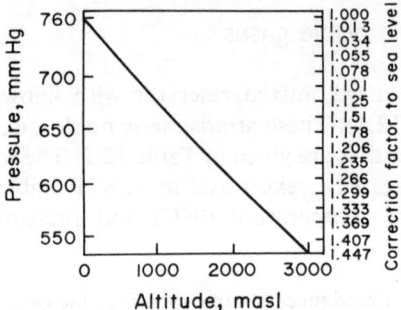

Fig. 12.2 Atmospheric pressure variations as a function of altitude (from the US Standard Atmosphere, 1962). Correction factors on the right axis serve to convert solubility values at sea level (Fig. 12.1) to values at desired altitude (dividing by the factor) or to normalize data at a given altitude to the corresponding value at 0 masl (multiplying by the factor). The last is needed to read intake (recharge) temperatures from Fig. 12.1, which is for sea level (i.e., 760 mm Hg) (Mazor, 1975).

proportional to their partial pressures in air. This, in turn, depends on the barometric pressure, which is linearly correlated to the altitude. The higher the altitude, the less noble gases dissolve in water. This relation is expressed in Fig. 12.2. The correction factor read on the right vertical axis in Fig. 12.2 is multiplied by a solubility value observed at a given altitude, in order to get the equivalent solubility (at the same ambient temperature) at sea level. Or, the solubility at a given temperature at sea level (Fig. 12.1) may be divided by the correction factor in order to get the solubility, at a given temperature, at a selected altitude. The need for these procedures will soon be explained, but let us first become familiar with them: what is the solubility of Kr at 18°C on a mountain 2500 masl? The answer is worked out in steps:

- The solubility of Kr at 18°C at sea level is (from Fig. 12.1): 7.5×10^{-8} cc STP/cc water.
- The correction factor for 2500 masl is (Fig. 12.2) 1.35, hence
- The solubility of Kr at 18°C and 2500 masl is:

$$\frac{7.5 \times 10^{-8}}{1.35} = 5.5 \times 10^{-8} \text{ cc STP/cc water}$$

The solubility of the noble gases also depends on a third parameter: the concentration of dissolved ions in the water. The data in Fig. 12.1 are for salt-free water. Sea water, for example, dissolves 30% less. This effect has only rarely to be regarded in groundwater tracing, as recharge water is in most cases relatively fresh.

12.3 Groundwater as a closed system for atmospheric noble gases

Recharge water equilibrates with air during its infiltration, and it dissolves atmospheric He, He, Ar, Kr and Xe in amounts defined by the local altitude and the ambient temperature. Once water enters the saturated zone of groundwater, no further additions of atmospheric gases are possible as the rock pores contain only water. Release of dissolved gases in the saturated zone is not possible, either, because of the hydrostatic pressure of the system. Because of these arguments it has been *postulated* that the noble gases are kept in closed system conditions in groundwater, and therefore one may measure the noble gas concentrations in water of a spring, or well, and calculate from them the ambient temperature at the recharge area.

A coastal spring was found to contain 12×10^{-9} cc STP Xe/cc water. What was the ambient temperature at the point of recharge (assuming the system remained closed)? From Fig. 12.1 it is seen that 12×10^{-9} cc STP Xe/cc water are dissolved in water that equilibrates with air at 12°C. Hence, if recharge would have been at sea level, the ambient temperature would have been 12°C.

A spring emerges at an altitude of 200 masl, the highest point in the potential recharge area is 400 m, and the average altitude of the recharge area is: $400 + 200/2 = 300$ m. In a water sample of this spring 10×10^{-9} cc STP Xe/cc water were found. What is the recharge temperature? Let us do the calculation in steps:

- The altitude correction factor for 300 m is read from Fig. 12.2 to be 1.05.
- Multiplying the observed Xe concentration by this factor, the equivalent concentration at sea level is obtained: $10 \times 10^{-9} \times 1.05 = 9.5 \times 10^{-9}$ cc STP/cc water.
- With this figure we can read the equivalent temperature, using Fig. 12.1. The result is 16°C.

In summary, it might be concluded that:

- The initial noble gas concentrations in recharged groundwater may be calculated from the ambient temperature and recharge altitude.
- The concentration of dissolved noble gases in groundwater (spring or well) may serve to calculate the intake ambient temperature, if the recharge altitude is known or assumed.

The scope of noble gas applications in groundwater studies is much greater than this, but first we have to check the basic assumption that groundwater provides closed system conditions for dissolved atmospheric noble gases.

12.4 Studies on retention of atmospheric noble gases in groundwater systems: cold springs

Groundwater samples from various locations in the world have been analysed for their noble gas isotopic composition and elemental concentrations. One such study was conducted by Herzberg and Mazor (1979) on four springs of simple hydrological structures (Fig. 12.3 and Table 12.3). The isotopic composition of the neon, argon, krypton, and

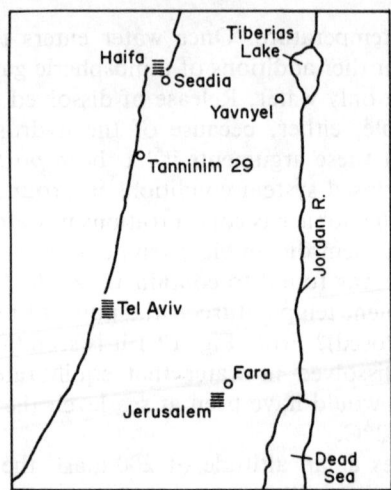

Fig. 12.3 Location of springs (o) in Israel, selected for noble gas retention studies (Table 12.3).

xenon was found to equal that of air, revealing the atmospheric origin. The noble gas concentrations are given in Table 12.4. These data were multiplied by the respective altitude correction factors (Fig. 12.2) to obtain the noble gas concentrations normalised to 1 atmosphere (sea level), as seen in Table 12.5. These values were then applied to calculate intake temperatures, which are given in Table 12.6 along with temperatures measured in the field.

Six points of temperature measurements seem to be especially informative in the hydrological cycle. These points are (Fig. 12.4):

- T_1 – the average ambient air temperature during the rainy season at the recharge location; obtained from meteorological surveys.
- T_2 – the average annual air temperature at the recharge location; obtained from meteorological studies.
- T_3 – the temperature at the base of the aerated zone, just above the water table. This temperature is deducible from the noble-gas concentrations.
- T_4 – the highest temperature reached along the groundwater cycle, presumably obtained at the deepest point of circulation.
- T_5 – the emergence temperature, observed at the spring or well head.
- T_6 – the ambient temperature at the site and time of sample collection.

T_4 is the only value we cannot measure, but if flow rates are high enough T_5 will be equal to T_4 or very little lower. T_5, the spring or well temperature, may thus be regarded as a minimum value of T_4.

The data in Table 12.6 reveal the following observations:

- Agreement between the intake temperatures calculated via Ar, Kr and Xe is ±1°C. This in itself indicates the systems were closed, as losses or gains would most probably be connected to solubility fractionations that would

Table 12.3 General data on springs selected for noble gas retention studies (Fig. 12.3) (from Herzberg and Mazor, 1979)

Name of spring	Sample no.	Date	Temperature (°C)	Altitude (m) Recharge area			Description
				maximum	average	minimum	
Saadia	OH-10	2.10.74	22.0	520	270	20	issues on the fault of Mount Carmel, on the contact with the Yzreal Plain
	OH-13	3.3.75	22.0				
Yavniel	H-6	14.9.74	20.5	300	225	150	a small spring in a basaltic region, draining two adjacent hills
Taninim 29	OH-16	19.3.75	23.5	900	450	10	one spring, out of several, issuing in the Kabara plain; seems to be recharged from the Shomron Mountains
Fara	OH-5	22.7.74	20.5	800	550	290	issues on the eastern slopes of the Judean Mountains

Table 12.4 Noble gas concentrations* (cc STP/cc water) in samples from springs in Israel (Table 12.3) (Herzberg and Mazor, 1979)

No.	Source	Date	Spring temp. (°C)	Ambient temp. (°C)	Ne ($\times 10^8$)	Ar ($\times 10^4$)	Kr ($\times 10^8$)	Xe ($\times 10^9$)
OH-10	Saadia	2.10.74	22.0	17	20.7 ± 1.0(4)	3.00 ± 0.10 (4)	6.49 ± 0.24 (4)	8.67 ± 0.60 (4)
OH-13	Saadia	4.3.75	22.0	12	20.4 ± 0.4(4)	2.99 ± 0.08 (4)	6.40 ± 0.25 (4)	8.56 ± 0.79 (4)
OH-6	Yavniel	14.9.74	20.5	21	21.3 ± 0.9(3)	3.01 ± 0.06 (3)	6.33 ± 0.33 (3)	8.65 ± 0.33 (3)
OH-16	Taninim 29	19.3.75	23.5	18	21.3 ± 0.3(3)	2.98 ± 0.04 (3)	6.31 ± 0.09 (3)	8.47 ± 0.31 (3)
OH-5	Fara	22.7.74	20.5	23	18.8 ± 0.4(3)	2.72 ± 0.11 (3)	7.59 ± 0.39 (3)	8.47 ± 0.31 (3)

* Errors cited are mean errors for several samples collected at the same time (their number is given in brackets). STP denotes 760 mm Hg and 0°C.

Table 12.5 The noble gas concentrations of Table 12.3, normalized to 1 atm (sea level) (cc STP/cc water)

No.	Source	Av. recharge altitude (m)	Altitude correction factor	Ne (× 10⁸)	Ar (× 10⁴)	Kr (× 10⁸)	Xe (× 10⁹)
OH-10	Saadia	270	1.030	21.3	3.09	6.68	8.93
OH-13	Saadia			21.0	3.08	6.59	8.82
OH-6	Yavniel	225	1.024	21.8	3.08	6.48	8.86
OH-16	Taninim 24	450	1.049	22.3	3.13	6.62	8.89
OH-5	Fara	550	1.068	20.1	2.90	6.23	8.11

Table 12.6 Intake (recharge) temperatures deduced from noble gases[a] and measured in the field[b] (°C)

No.	Spring	Date	Recharge area		Noble gases, T_3				Sampling site	
			$T_1{}^c$ average rainy season	$T_2{}^c$ average	Ar	Kr	Xe	averaged[d]	T_5 spring	T_6 ambient air
OH-10	Saadia	2.10.74	10	19	21	21	21	21 ± 1	22.0	17
OH-13	Saadia	4.3.74	10	19	21	22	21	21 ± 1	22.0	12
OH-6	Yavniel	14.9.74	14	21	21	23	21	21 ± 2	20.5	21
OH-16	Taninim 19	19.3.75	14	19	21	22	22	21 ± 1	23.5	23
OH-5	Fara	22.7.74	10	16	23	24	23	23 ± 1	20.5	23

[a] Applying the data of Table 12.5 and the solubility curves (Fig. 12.1).
[b] T_1–T_6 are defined in the text and in Fig. 12.4.
[c] From various meteorological sources.
[d] The cited errors were calculated from the errors of the individual runs as given in Table 12.4.

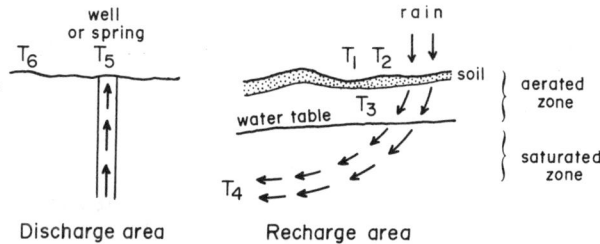

Fig. 12.4 Definition of temperatures relevant to hydrological studies: T1, average temperature in the rainy season at recharge area; T2, average annual temperature at recharge area; T3, temperature at the base of aerated zone, above the water table (deduced from Ar, Kr and Xe concentrations); T4, maximum temperature reached at the deepest point of the water path; T5, observed spring or well temperature at time of sampling; T6, ambient air temperature at the time of sampling (from Herzberg and Mazor, 1979).

introduce disagreements between the various calculated intake temperatures.
- In the first three cases (Saadia, Yavniel, and Taninim 29) the noble gas deduced temperatures (T_3) are equal to, or 2°C higher than, than the average temperature. In contrast, they are higher by about 10°C than the average temperature of the cold rainy season. Hence, it seems well established that the temperature documented by the noble gases, during equilibration with air at the base of the aerated zone, equals the local average annual temperature. This makes sense because infiltrating recharge water often descends at rates of a few metres per year or less. Hence, infiltrating water is delayed in the aerated zone, adopting the average temperature prevailing underground. This topic will be further discussed in the next section.
- In the first three samples of Table 12.6 the noble-gas deduced temperature, T_3, differs from the ambient air temperature at the time of sample collection, T_6, revealing that no equilibration at the point of emergence took place (in agreement with the previous paragraph). The last spring (Fara), however, shows $T_3 = T_6$. This water issues below a slope of tallus, and equilibration of noble gases could take place prior to sample collection.

Good agreements, obtained in an ever-growing body of measured cold groundwaters, between noble-gas deduced intake temperature and average annual temperature at recharge area, *prove* the validity of the original assumption that groundwater is a closed system for the dissolved atmospheric noble gases.

12.5 Further checks on atmospheric noble gas retention: warm springs

The closed nature of cold groundwater systems to atmospheric noble gases may be further checked in warm springs, i.e. in waters that percolate to 1-2 km depth and are heated by local (normal) heat gradients. The first study of this type was done on warm springs and wells (up to 60°C) of the Jordan Rift Valley (Mazor, 1972). The *isotopic composition* of the encountered dissolved Ne, Ar, Kr and Xe were found to be atmospheric, i.e. very close (in the range of the analytical accuracy) to the values given in Table 12.2. The atmospheric noble gases are presented in Fig. 12.5 in a fingerprint diagram. An overall similarity of the line patterns to those of air-saturated water at 15-25°C is clear. Thus, the *relative abundances* of the four noble gases, Ne, Ar, Kr, and Xe again indicate they originated from air dissolved in the recharge water. The fingerprint lines in Fig. 12.5 are parallel to each other but vary in their height on the concentration axis - what might be the explanation? Differences in the recharge areas and, hence, in the initial noble gas concentrations, due to differences in altitude (temperature and pressure). In one case, the Tiberias Hot Springs, the initial salinity played a significant role. Let us have a closer look at the Tiberias Hot Springs case: it is lowest in Fig. 12.5, revealing lowest noble gas

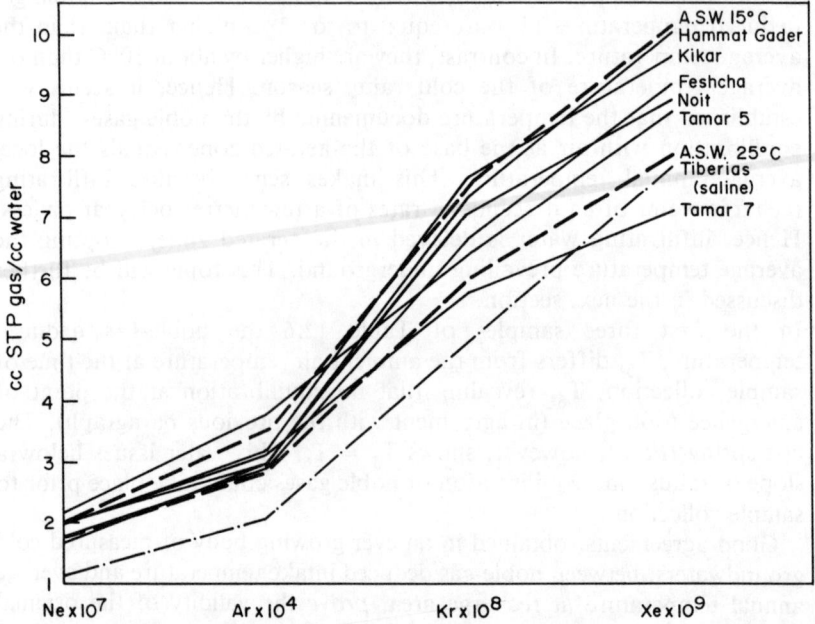

Fig. 12.5 Noble gases dissolved in thermal waters in the Jordan Rift Valley (data from Mazor, 1972). The non-saline waters lie between the values of air-saturated water (ASW) at 15°C and 25°C, demonstrating the meteoric (atmospheric) origin and the closed-system conditions that prevail in the ground. The saline Tiberias hot water is suggested to have originated by sea water entrapment, a hypothesis supported by the lower noble gas concentration (close to sea water saturation at about 15°C) (Mazor, 1972).

concentrations. However, it has a salt concentration almost equal to that of sea water and, based on geochemical and hydrological arguments, it has been suggested that this water originated from infiltration of sea water when the Mediterranean Sea invaded the Rift Valley for a short period (Mazor and Mero, 1969b). If so, the initial noble gas concentration was as in sea water, i.e. 30% less than in fresh water under similar conditions. Hence, to normalize the Tiberias Hot Spring line in Fig. 12.5, and compare it with the fresh water samples, the values have to be increased by 30%. If we do this, the Tiberias Hot Spring line falls right in the middle of the other Rift Valley warm waters, close to the air saturated water (ASW) at 15°C, agreeing with closed system conditions. The noble gases thus provide independent support for the origin of the Tiberian Hot Spring from sea water.

As noble gas concentrations are obtained, the hydrochemist is always confronted with the need to establish whether the data reflect the indigenous noble gas concentrations, or whether re-equilibration with the atmosphere occurred prior to sampling. The problem is acute in springs: the samples are collected in the spring as deep as possible below the water surface, but did the water re-equilibrate while flowing below gravel, for

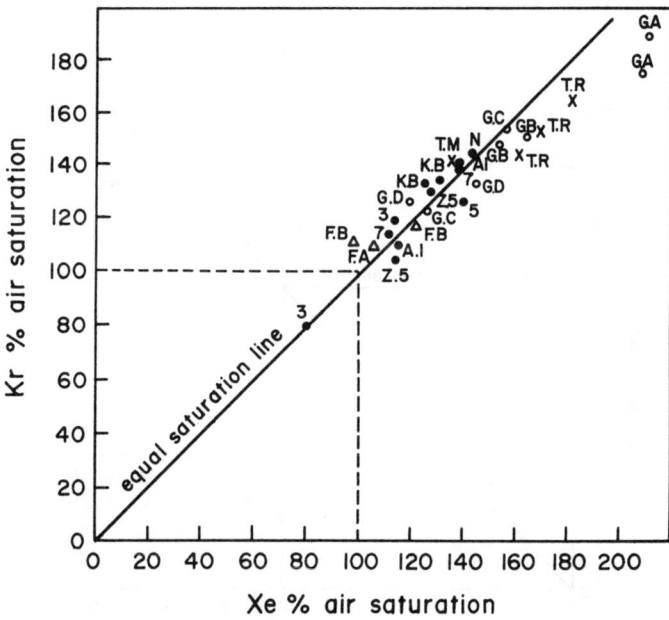

Fig. 12.6 Percentage of air saturation, or retention, of Kr and Xe for the Rift Valley waters. The values were obtained by dividing the measured amounts by those expected for water equilibrated with air at the temperature at which each sample was collected. All samples, except one, were found air-supersaturated, indicating that the rare gases were retained under closed system conditions. Differences in duplicate samples are attributed to gas losses to the atmosphere prior to sampling (Mazor, 1975).

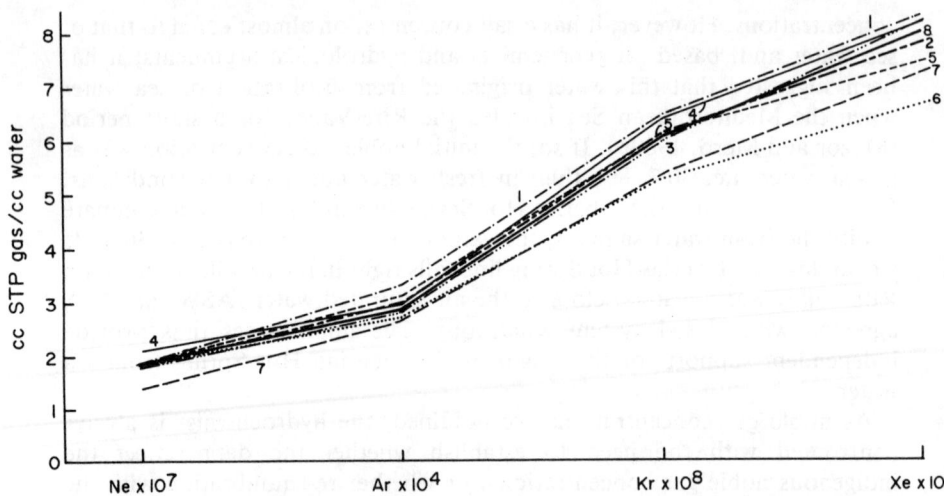

Fig. 12.7 Noble gas pattern of atmospheric noble gases in warm springs in Swaziland. The isotopic compositions were found to be atmospheric and the abundance patterns, so similar to ASW at 25°C, confirm this. The variations in concentrations are caused by different recharge altitudes (Mazor et al., 1974).

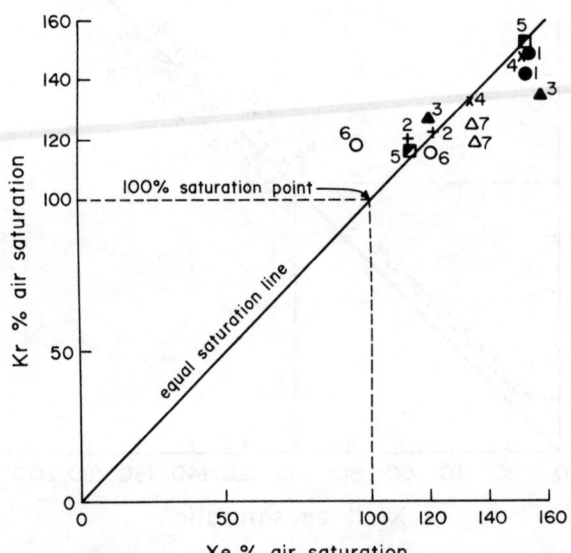

Fig. 12.8 Percentage air saturation, or retention, of krypton and xenon for warm springs in Swaziland (some spring numbers as in Fig. 12.7). All samples were oversaturated at the temperature of emergence (35°C–52°C), indicating the systems were closed (Mazor et al., 1974).

example, or did the water circulate in the spring's outlet? An answer may be obtained by comparing the observed noble gas concentration with the concentration calculated for equilibrium with air at the emergence temperature of the spring.

An example: a spring emerges at 34°C and is found to contain 3.8×10^{-4} cc STP Ar/cc water. Is it oversaturated, i.e. does it contain more gas than expected from air-equilibration at the emergence temperature? The air-water solubility at 34°C is seen in Fig. 12.1 to be 2.5×10^{-4} cc STP Ar/cc water. Hence, the observed value of 3.8×10^{-4} cc STP/cc water is oversaturated by:

$$\frac{3.8 \times 10^{-4}}{2.5 \times 10^{-4}} \times 100 = 152\%$$

of the equilibration value at the temperature of emergence (the observed value indicates a recharge temperature of 10°C - is this right?).

The observed Kr and Xe concentrations in each water source in the Rift Valley study were converted in this way to percentage air saturation in Fig. 12.6. It is seen that the samples contain more than 100% air saturation, i.e. they are oversaturated, indicating that re-equilibration subsequent to the heating did not occur. This diagram shows the importance of duplicate or triplicate sampling. The one low value of sample 3 is clearly wrong due to gas loss during sampling and the higher sample 3 values are more representative (and not the average).

Warm spring waters circulate deep, and their ages are relatively high. Hence they are a rigorous check for the prevailance of closed-system conditions for the atmospheric noble gases in the saturated zone. Figures 12.7 and 12.8 show results for another case, in Swaziland.

12.6 Computation of paleotemperatures

Old groundwaters contain dissolved noble gases in concentrations defined by the ambient temperature that prevailed at the time of recharge. Coupling of groundwater dating methods, e.g. ^{14}C (Chapter 11), with intake temperatures calculated from observed Ar, Kr and Xe concentrations (section 12.3), may provide relevant paleotemperatures (Mazor, 1972).

An example of a noble gas paleotemperature study has been reported by Andrews and Lee (1979) from the Bunter Sandstone aquifer, eastern England. The noble gases were studied as part of an intensive hydrochemical study, discussed in section 11.9. Table 12.7 contains the noble gas concentrations in three sections of the studied groundwater system: group A water-recharge zone, phreatic aquifer; group B water, confined section, close to recharge zone; group C water, confined, distant from recharge area (section 11.9). The observed noble gas concentrations served to calculate intake temperatures. The results: $10 \pm 1°C$ for group A, $9 \pm 1°C$ for group B, and $4 \pm 1°C$ for group C. The paleotemperatures are plotted versus emergence temperature (indicating depth) in Fig. 12.9.

Table 12.7 Noble gases (cc STP/cc water) of groundwaters in the Triassic sandstone aquifer in eastern England (from Andrews and Lee, 1979), deduced intake temperatures (read from solubility curves. Fig. 12.1) and radiogenic He*

Group	No	Well	He×10⁻⁸	Ne×10⁻⁷	AR×10⁻⁴	Kr×10⁻⁸	Rad. He* 10⁻⁸	Deduced intake temp. (°C) Ne	Ar	Kr	Average
A	5	Elkesley, No.6	4.0	2.09	4.02	8.65	−0.8	10	9	12	10 ± 1
A	9	Elkesley, No.5	4.3	2.09	4.02	8.68	−0.5	10	9	12	10 ± 1
A	12	Elkesley, Clark's No.1	7.4	2.09	3.87	8.69	2.6	10	10	12	11 ± 1
A	6	Far Baulker, No.3	4.3	2.09	4.05	8.78	−0.5	10	9	12	10 ± 1
A	1	Amen Corner, No.1	4.2	2.09	3.83	9.02	−0.6	10	11	11	11 ± 1
A	3	Boughton, No.2	6.2	2.01	4.07	8.71	1.5	15	8	12	12 ± 2
A	10	Retford, Clark's No.2	6.5	2.07	3.94	9.14	1.7	11	9	11	10 ± 1
A	8	Ompton, No.2	7.8	2.12	4.16	9.12	2.9	9	7	11	9 ± 2
A	4	Fransfield	5.1	2.09	4.09	8.70	0.3	10	8	12	10 ± 2
A	7	Rufford, No.3	4.9	2.02	3.89	8.74	−0.2	(15)	10	11	11 ± 1
A	11	Retford, Orsdall No.1	3.5	2.09	4.04	8.64	−1.4	8	7	12	9 ± 2
A	13	Everton, No.1	4.3	2.07	3.76	8.84	−0.5	11	11	11	11 ± 1
A	14	Everton, No.3	4.2	2.05	4.28	9.13	−0.6	10	11	11	11 ± 1
A	15	Retford, Whisker Hill	4.6	2.08	3.89	8.80	−0.2	10	10	12	11 ± 1
B	16	Halam, No.1	14.1	2.23	4.15	8.82 9.2		6	7	12	8 ± 2
B	17	Markham Clinton, No.1	4.3	2.09	3.99	8.74	−0.5	10	9	12	10 ± 1
B	21	Egmanton, B.P.	10.8	2.11	4.07	9.13	5.9	9	8	11	9 ± 1
B	20	Caunton	3.6	1.84	3.65	7.93	−1.1	(25)	12	15	9 ±
B	19	Retford, Grove No.2	8.4	2.14	4.19	9.13	3.6	5	7	11	8 ± 2
C	26	Gainsborough, B.P.	23.7	2.20	4.58	9.85	18.7	5	3	8	5 ± 2
C	24	Newark, British Gypsum	21.9	2.29	4.86	10.63	16.9	3	2	5	3 ± 1
C	25	Newark, Castle Brewery	25.9	2.27	4.78	10.50	20.9	4	2	5	4 ± 1
C	30	Gainsborough, Humble Carr	27.2	2.18	4.49	10.10	22.2	5	4	7	5 ± 1
C	28	Newton, No.2	17.3	2.27	4.72	10.54	12.3	4	3	5	4 ± 1
C	31	Gainsborough, No.3	34.3	2.20	4.52	10.28	29.3	5	4	6	5 ± 1
C	22	Rampton	19.3	2.27	4.68	10.48	14.3	4	3	5	4 ± 1
C	32	South Scarle	17.6	2.20	4.51	9.96	12.6	5	4	8	6 ± 2

* Observed He minus atmospheric He, read from the solutility curve (Fig. 12.1), applying the average deduced intake temperature (last column of present table).

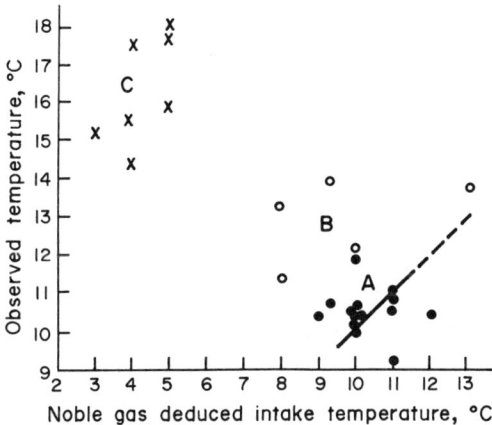

Fig. 12.9 Noble gas deduced intake temperatures versus observed emergence temperatures (Table 12.7) in wells of the Bunter sandstone aquifer, eastern England. The dashed line is an equal temperature line, i.e. noble gas temperature equals observed temperature. Group A data are close to this line. Group B reveals slightly colder paleotemperatures (8°-10°C) and group C reveals significantly colder values (3°-5°C).

Other examples of noble gas paleotemperatures have been reported by Rudolph *et al.* (1983). They calculated temperatures as low as 1-2°C 22 000 years ago (dated by ^{14}C) in Germany, accompanied by relatively negative δD and δ^{18}O values and measurable excess of radiogenic He.

12.7 Calculation of depth of circulation and location of recharge zone

Infiltrating water is, in many cases, delayed in the aerated zone long enough to equilibrate to the ambient temperature. In shallow depths, e.g. less than 10-15 m, this will be close to the local average annual temperature. Daily temperature fluctuations are damped at less than 0.5 m (Fig. 12.10) and seasonal fluctuations are damped to a number of metres.

If the temperature of a spring, or well, is higher than the noble-gas deduced temperature (T5 > T3 in Fig. 12.4), then the additional temperature reflects heating by the local geothermal gradient. If this gradient is known, then dividing the temperature gain by the gradient will provide the depth to which the water circulated in the saturated zone.

Example. The water of the Taninim 29 spring is seen in Table 12.6 to emerge at 23.5°C, wheras the noble-gas deduced temperature at the base of the aerated zone is 21 ± 1°C. How can this difference be explained? The additional 2.5°C resulted from equilibration at the aquifer temperature. The local heat gradient is not known, so let us use the common value of

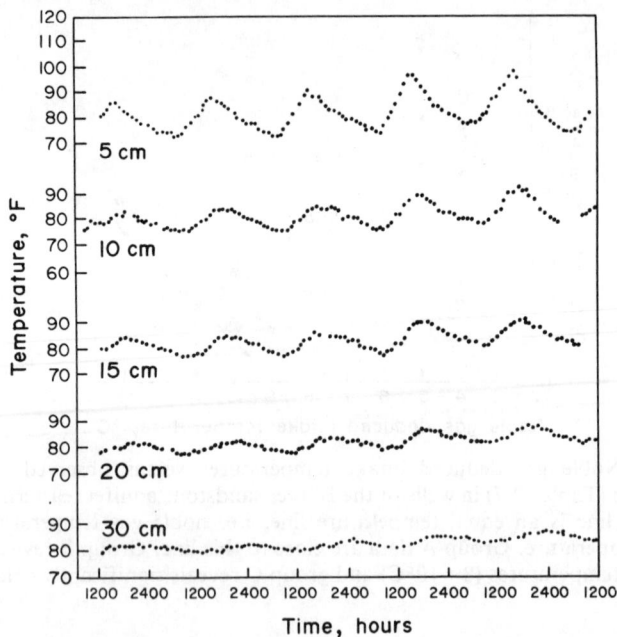

Fig. 12.10 Daily temperatures in soil profiles, measured in 1970 at the Santa Rita Experimental Range, Tucson, Arizona, USA. The daily fluctuations are seen to be practically damped at 30 cm (after Foster and Fogel, 1973).

3°C/100 m. Accordingly, the water circulated to a depth of 80 m or more in the saturated zone. (It might be more because partial cooling during ascent could occur.)

Locating recharge areas. A well in a plain of 100 masl provides fresh water at 24°C. The question came up whether it is recharged by local rain, or from snow melt on the mountains 20 km distant. To resolve this problem the noble gases were measured and the following concentrations were found: 2.1×10^{-7} cc STP Ne/cc water and 3.9×10^{-4} cc STP Ar/cc water. Where is the recharge location? The neon curve in Fig. 12.1 is seen to vary little with temperature, and therefore it is of little use in intake-temperature circulation. The argon curve is, in contrast, sensitive to temperature changes. An intake (base of aerated zone) temperature of 8°C is obtained. Hence, $24 - 8 = 16°C$ is the temperature gain obtained by circulation, reflecting a distant and high intake – supporting recharge from the mountain. If so, a respective altitude correction should be applied to the raw noble-gas data, resulting in an intake temperature lower than 8°C, further supporting mountain (and snow melt) recharge.

The last example demonstrates the potential use of the atmospheric noble gases in identifying a recharge location. Combined with stable isotope measurements of hydrogen and oxygen in water, i.e. δD and $\delta^{18}O$ values (section 9.8) and then, applying the respective altitude correction to the noble gas data, one may calculate the intake temperature, thus checking the validity of the suggested recharge location. Furthermore, the difference between observed emergence temperature and noble-gas deduced intake temperature will give the minimum depth of circulation and, thus, tell whether the water moves underground in a short line or first descends deep and subsequently ascends.

12.8 Identifying karstic recharge

In Table 12.8, data for two karstic springs from northern Israel and one from eastern Switzerland are given. The table includes:

- Ne, Ar, Kr and Xe concentrations,
- the noble-gas concentration in air saturated water at the respective recharge temperatures and altitudes, and
- dividing (a) by (b) and multiplying by 100, the percentage concentration, relative to the expected air saturated water at recharge point, is obtained.

The values (Fig. 12.11) are significantly different from those seen so far:

- Ne concentrations are 144–172%. Thus, *excess atmospheric* Ne (recognized as atmospheric by its isotopic composition) is present.
- The other noble gases are also present in concentrations exceeding 100%, but the excess decreases towards Xe, and in other words, *the percentage retention* pattern is:

Ne > Ar > Kr > Xe

Table 12.8 Stages in the calculation of percentage noble gas retention in karstic water samples

Data	Karstic spring	Ne	Ar	Kr	Xe
Measured values	Dan, Israel	28.8	36700	7.85	1.02
10^{-8} cc STP/cc water[a]	Banias, Israel	24.1	33800	6.97	0.97
	Source de la Doux, Switzerland	28.6	39600	8.81	1.13
ASW, 14°C, 1700 m[b]	recharge of Dan and Banias springs	16.7	28000	6.70	0.96
ASW, 10°C, 1000 m	recharge of Source de la Doux	18.3	32000	7.00	1.09
% retention, relative	Dan	172	131	116	106
to AWS at recharge point[c]	Banias	144	121	104	101
	Source de la Doux	156	124	116	104

[a] Data from Herzberg and Mazor (1979).
[b] The values, in 10^{-8} cc STP/cc water, were calculated from the relevant solubility values, divided by the corresponding altitude correction factor.
[c] *(a/b)* × 100.

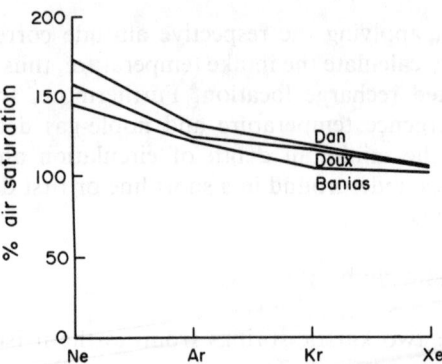

Fig. 12.11 Percentage air saturation for karstic springs (Table 12.8). Values are above 100% and the lines reveal a 'reversed' pattern of Ne > Ar > Kr > Xe.

The explanation is that free air has been siphoned into the karstic water, in addition to the initial dissolved air. The solubility of the noble gases is smallest for He and increases towards Xe. This should not be confused with Fig. 12.1, which was obtained by taking the solubilities of each noble gas and multiplying by their abundance in air (Table 12.1). (One may obtain the noble gas solubilities for a selected temperature by dividing the respective values in Fig. 12.1 by the abundance values of Table 12.1.) Free air is relatively richer in the less soluble gases, and therefore its addition to already saturated water causes in karstic springs (a) the excesses over 100% saturation, and (b) the excess pattern of Ne > Ar > Kr > Xe.

The excess atmospheric air was observed to be negatively correlated to spring discharges, reflecting that in the low-flow season larger parts of the system contain air, and when the water flows through narrowing conduits the air is sucked in. In the high-flow season most of a karstic system is filled with water and less air can be siphoned. Excess air in groundwater has been reported by Heaton and Vogel (1981) for cases of special soil conditions observed in South Africa. Hence, periodical noble gas measurements are desired: a negative correlation with discharge will establish the karstic nature of a studied spring or well.

12.9 Radiogenic He and its use as an age indicator

The radioactive decay of uranium and thorium results in the formation of a series of isotopes that are radioactive and go on disintegrating to form stable lead isotopes. These radioactive disintegrations are accompanied by the emissions of ^{4}He atoms. Three such radioactive series exist:

$$^{238}U \rightarrow {}^{206}Pb + 8\, {}^4He$$
$$^{235}U \rightarrow {}^{207}Pb + 7\, {}^4He$$
$$^{232}U \rightarrow {}^{208}Pb + 6\, {}^4He$$

Common rocks contain uranium and thorium in ppm (parts per million) concentrations. Thus, radiogenic helium is constantly formed in rocks. This helium is partially released from the crystal lattices of rocks, and is dissolved in the associated water. The production of helium is constant in any given set of rocks, and the amount dissolved in stored groundwater is proportional to the length of water storage. This point is demonstrated by the comparison of groundwater ages deduced from ^{14}C and dissolved helium in three groups of groundwater of the Bunter sandstone, eastern England (Fig. 11.26). A linear correlation is observed, supporting the ^{14}C deduced ages, and demonstrating the applicability of radiogenic helium as an age indicator. The term *age indicator* is used with regards to radiogenic helium rather than *dating method*. This is to emphasize the semi-quantitative, yet hydrologically valuable, nature of the helium dating. The amount of ^4He stored in groundwater is controlled by several factors:

- Concentration of uranium and thorium in aquifer rocks.
- Water : rock ratio.
- The efficiency with which He is emanated from the rocks.
- Age of the groundwater.

Of these variables, age seems to be the dominant one.

The ^{14}C dating method is based on a known concentration of ^{14}C in the initial water, and depletion of ^{14}C with age. Thus, the older the water gets, the more difficult it is to measure the ^{14}C that is left in it. The ^4He method has the opposite pattern: the initial concentration of radiogenic helium is zero in groundwater, but it accumulates with age and the older the water is, the easier it is to measure the concentration of the radiogenic ^4He. Thus, the two dating methods are complementary. A practical limit of ^{14}C dating of groundwater is about 25 000 y, whearas the range of radiogenic ^4He is 10^4–10^8 y.

The concentration of helium in air-saturated water is at most 5×10^{-8} cc STP/cc water, as seen in Fig. 12.1. Any excess above this concentration is of non-atmospheric helium. In terrains located on stable parts of the crustal plates, radiogenic helium is the sole non-atmospheric source. However, along plate margins and volcanic systems, mantle He is added to groundwater as well. These mantle He contributions are noticeable by their relatively high $^3He/^4He$ ratios, and rarely reach 50% of the observed helium. In addition, mantle helium is mainly added by release from mantle-produced rocks. This release is in itself time correlated, and interferes only slightly, if at all, with the application of He as an age indicator.

In summary, radiogenic He complements the atmospheric noble gases as a useful hydrochemical parameter (Mazor and Bosch, in press).

12.10 Sample collection for noble gas measurements, and contact with relevant laboratories

Noble gases are measured with special mass spectrometers. The size of samples needed is 5-10 cc. The major difficulty in noble gas work is the need to absolutely avoid air contamination: a bubble of air 1/1000 of the volume of the sample contains as much noble gas as the whole amount dissolved in the water sample. Two major techniques are in use to sample for noble gas measurements:

- filling water in glass tubes with special high vacuum glass stopcocks, or
- filling copper tubes, closed at their ends with special clamps.

The sampling method should be talked over with the laboratory staff, who have also to provide the collection vessels and detailed sampling instructions. As the sample collection has to be done with care, it is recommended that laboratory personnel do the sample collections if possible.

The number of laboratories that measure noble gases in water samples is slowly, but steadily, growing. The know-how is actually at hand in most laboratories that do potassium-argon dating of rocks. It seems that by demand from field hydrochemists, more laboratories will be willing to enter the domain of noble gas hydrology, and noble gas measurements will become as routine as δD, $\delta O^{18}O$, tritium, or carbon-14 measurements.

Note: a general text on terrestrial noble gases is that by Ozima and Podosek (1983).

13 MONITORING OF CONTAMINANTS

13.1 Scope of the problem

Large-scale human operations carried out on the surface of the Earth include ever-increasing intervention in groundwater systems. All uses of water result in increase of dissolved components, all used water has to be disposed of somewhere, and all disposed water reaches the existing water reservoirs: lakes, rivers, groundwater and the ocean. Hence, an intrinsic part of water use is water contamination, i.e. addition of components that were not there before. Water quality control is, thus, an integral part of water production and consumption and, in the same way, water monitoring and tracing of pollutants is an integral part of hydrochemists' field work.

Monitoring of contaminants often has the nature of trouble shooting, but in effect, each contamination case study also has a scientific aspect, because *each polluting source serves as a large-scale tracing experiment*. Chapter 6 was devoted to the establishment of hydraulic interconnections – the basic premise that underlies most groundwater models. Pollutants are useful tracers in this regard.

The use of artificial tracers, introduced into groundwater systems for direct tracing of hydraulic interconnections, is limited to wells that are tens of metres apart, and in karstic systems artificial tracing may extend for a few kilometres. For larger-scale experiments enormous quantities of tracers are needed. These are costly, and are never authorized by the local health authorities. Pollution by a land fill, waste dump, sewage disposal, factory effluents, or pesticide storage – all are threatening, and financial resources are available to deal with them. It is our challenge to use such cases as large-scale tracing experiments, and also to *use each catastrophe to improve our understanding of the hydraulic underground plumbing*.

In essence, contamination monitoring and pollutant tracing are special cases of the general topic of water mixings. Hence the emphasis put on the detection of mixing of different water types, identification of end members, and calculation of mixing percentages (sections 6.4, 6.6, and 6.7). The mixing in contaminated water systems is twofold:

- Introduction of the contaminants into uncontaminated water.
- Introduction of contaminants carried by water that differs from the water of the threatened system.

Examples: irrigation water is partially evaporated and therefore becomes enriched in deuterium and oxygen-18, providing a signature that differs from the local groundwater into which the irrigation water carries fertilizers and pesticides. Water used by industry comes often from rivers, or surface reservoirs, with atmospheric tritium concentrations, whereas threatened groundwater may be of a pre-1952 age and devoid of measurable tritium.

In the deciphering of contamination all the tools of the hydrochemist come into play: many parameters have to be measured, repeated sampling is needed to understand the dynamics of the systems, and large amounts of data have to be processed. Based on the observations, a conceptual model is constructed and this model has to be checked by quantitative (mathematical) modelling and also by further checking of predicted consequences.

13.2 Detection and monitoring of pollutants: some basic rules

Pre-pollution database. Let us begin the discussion with a question: 240 mg/l NO_3 have been detected in a farm well, far above the permissible concentration for domestic water. What should be done? Two steps are needed:

- Notifying the local people of the finding.
- Searching for the source of the contaminant.

The first step in this search is to establish the time when the pollutant arrived, and, in the case of the nitrate, is it outside contamination or a local natural phenomenon?

When a pollutant is detected, several contaminating candidates may be suspected: fertilization by farmers A, B and/or C, three different cattle stations, or the passage to deeper ploughing machines that has caused aeration and nitrification in a deeper soil layer. Knowledge of the time of pollutant arrival may eliminate suspected sources that arrived at later dates, and narrow the list of possible sources of contamination. Hence the need for historic data (section 7.5). These may be gathered from local authority archives, obtained from the well owner, or found in published reports. The recommendation to collect the historical data as a basic part of every pollution investigation sounds trivial, but in practice it is only partially followed, due to lack of awareness of this point. (It is easier to accuse the nearest suspected polluter and let him defend himself.)

The need for a pre-pollution database is a major incentive for regional hydrochemical surveys that have to be performed periodically.

Identification of specific 'labels' in potential contamination sources. Work on potential pollution sources necessitates thorough knowledge of their composition. Maximum parameters should be measured on every industrial effluent that leaves a factory and the fluid that is formed in each landfill. The parameters to be measured may be temperature, pH, major dissolved ions, trace elementals – mainly metals, metal compounds, organic compounds and isotopic compositions of hydrogen, oxygen, carbon, sulphur, or nitrogen. To perform such analyses, special laboratories have to be contacted. In certain cases the main product may serve as the label, and in other cases the labels are supplied by accompanying compounds, which in themselves may not be poisonous.

Movement of pollutants in the aerated zone. Most pollutants are released on the ground and have to pass the aerated zone before they arrive at the saturated groundwater zone. The nature of the aerated zone must be well known in each case. Aerated zones with dominating flow in pores may retard waste waters and cause:

- Biological decomposition of part of the toxic components.
- Absorption of toxic ions on clay particles.

These protections are absent in areas of conduit-dominated recharge. Hence, knowledge of the flow regime in the aerated zone may help to assess the influence of potential contaminating agents.

The mode of fluid flow in the aerated zone may be changed by the hydraulic aspects of fluid waste disposal. Constant release of large amounts of fluids may cause a local rise in the hydraulic head. Coupled with chemical fluid-rock interactions, this may form new high-conducting conduits that may lead the contaminants directly into the saturated zone. In other cases, fine particles that come with the contaminating fluids, may clog pores in the aerated zone and reduce through flow.

Importance of very sensitive monitoring techniques. The propagation of pollutants takes time, and the arrival in springs and wells is often gradual. In optimal set-ups the arrival of contaminating fluids may be detected by changes in the concentration of one or several parameters that are measurable but not toxic. In such cases preventive measures may be operated in due time. Hence the emphasis on most sensitive measuring techniques – either of *in situ* measuring probes or of laboratory procedures conducted on collected water samples.

Importance of extensive hydrochemical studies. Hydrochemical surveys that encompass many wells and springs in extended areas may spot water sources of exceptional compositions, possibly caused by manmade contamination. Such surveys may detect contamination long before any complaints are expressed by water consumers. The next step in dealing with water sources that reveal suspected contamination is comparison of the composition with historical data, as discussed above.

13.3 Groundwater pollution case studies

A large number of groundwater pollution cases are never published, for various reasons. This is especially true for severe cases. Yet, an ever-growing number of case studies is published. Every hydrochemist should read as many of these papers as possible, as each provides examples of possible scenarios and possible modes of diagnoses and remedies. A selection of pollution case studies is reviewed in the following sections.

Nitrate contamination, detected by depth profiles

Pickens *et al.* (1978) worked on nitrate contamination in wells drilled in a shallow sandy aquifer. A contaminated nitrate plume was detected in carefully sampled depth profiles (section 7.3), as shown in Fig. 13.1. The contamination was attributed to extensive use of nitrate fertilizers.

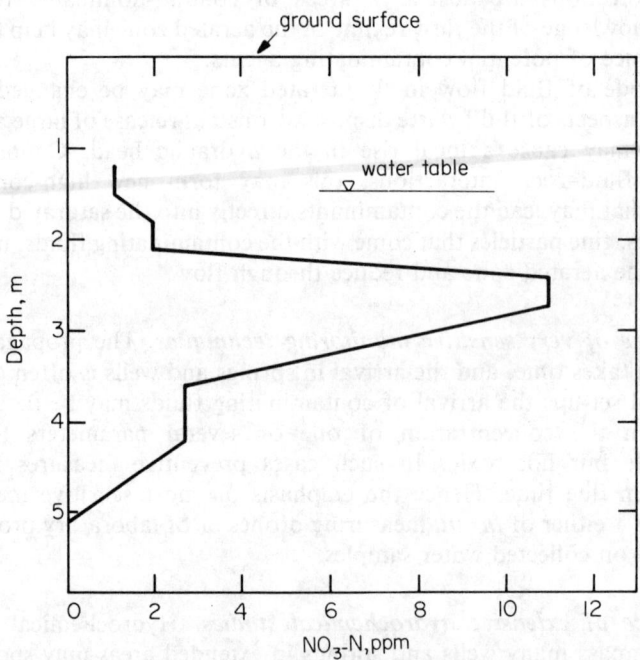

Fig. 13.1 Nitrate (expressed in ppm N) in a depth profile in a piezometer (from Pickens *et al.*, 1978). A contaminated water plume is seen at a depth of 3 m.

Discussion

- The arrival of the pollutant was detected in a very early stage – 12 ppm of nitrate being alarming in this case, but not threatening. The time was gained for remedial measures.
- The nitrate signal would not be seen on a pumped sample that would lower the nitrate concentration by its averaging effect, hence the advantage of depth profiles was demonstrated in this case.
- The nitrate peak indicates higher water conductance in the 2.5–3.0 zone.

The study reached the two aims outlined at the beginning of the present chapter: detection of the contaminating agent, and gaining an insight into the dynamics of the natural system.

Oxidation and reduction zones in groundwater profiles and their bearing on nitrate contamination

Andersen et al. (1980) worked on nitrate distribution in the Karup basin, Denmark. Results of depth profiles in two boreholes are shown in Fig. 13.2. 13.2.

Discussion. NO_3 is seen to decrease at a water depth of 8 m (15 m below surface), whereas Fe is low at this depth and rises beneath. This indicates existence of an oxidizing zone in the upper 8 m of water and a reducing zone below. The nitrate was reduced by ferrous iron to N_2, which was expelled into the atmosphere. The tritium data were most informative: values up to 200 TU were detected in the NO_3 contaminated upper water zone, indicating post-1963 (man-made tritium peak) recharge (the study was conducted in 1976). Thus, fast-arriving recharge was demonstrated, supporting the hypothesis of contamination by locally used fertilizers. The investigators determined the soil capacity to contribute ferrous iron to the reaction. This value is needed to assess for how long one may rely on this mode of nitrate decomposition, and to predict when nitrification will become a problem, if use of excess fertilizers continues.

Water salinization due to removal of vegetation

Borman and Likens (1970) reported changes in the composition of water flowing from a watershed, caused by removal of vegetation at the Hubbard Brook Experimental Forest, New Hampshire, USA (Fig. 13.3).

Discussion. The time of experimental removal of vegetation is indicated with arrows in Fig. 13.3. A distinct rise in nitrate and calcium was noticed, while no rise was noticed in an adjacent non-disturbed watershed. Thus, the increase in nitrate was clearly related to the removal of vegetation, that

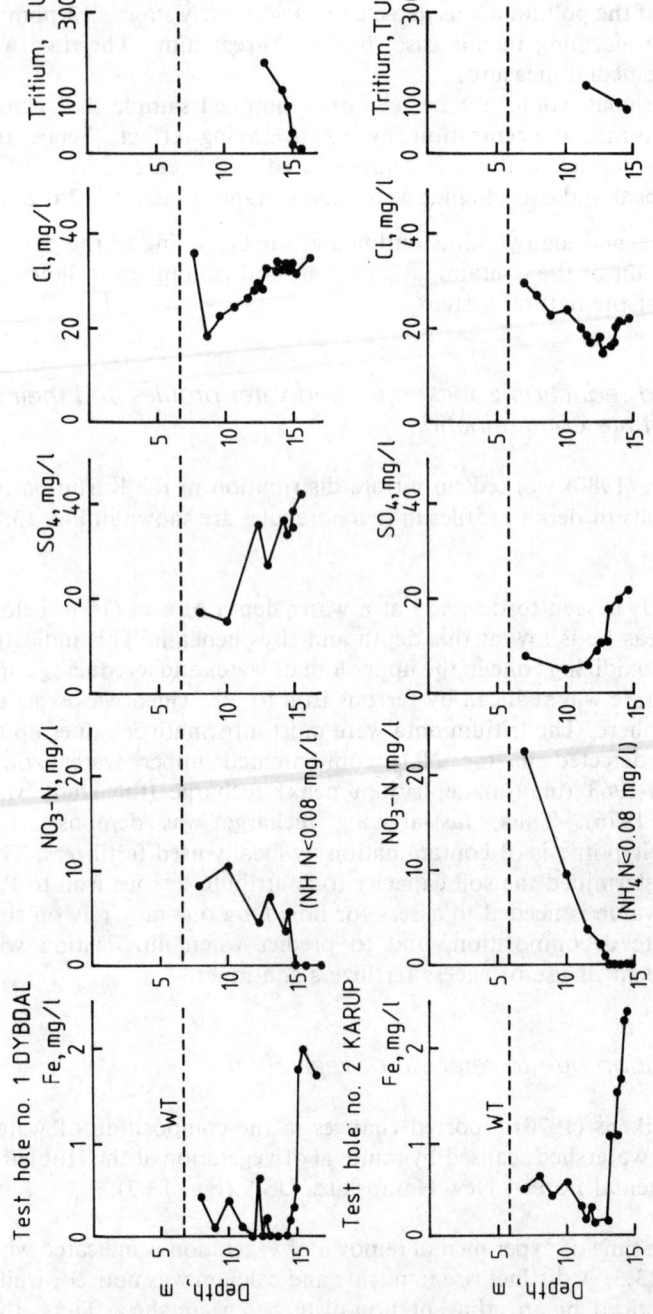

Fig. 13.2 Depth profiles of Fe, No₃, SO₄, Cl and tritium in two test holes, Karup Basin, Denmark (Andersen et al., 1980). A sudden change is observed at a depth of 15 m (8 m below water table), explained as indicating occurrence of an oxidizing regime above 15 m and a reducing regime below (see text).

lowered the denitrification effect. The increase in nitrate and calcium was abrupt, and occurred 5 months after the vegetation clearing, demonstrating a piston flow recharge. Studies of this nature are relevant to the influence of deforestation and vegetation stripping by coal mining. Such operations may adversely affect the quality of local surface and groundwater. Thus, such operations should be accompanied by hydrochemical monitoring before, during, and after termination of the intervening operations.

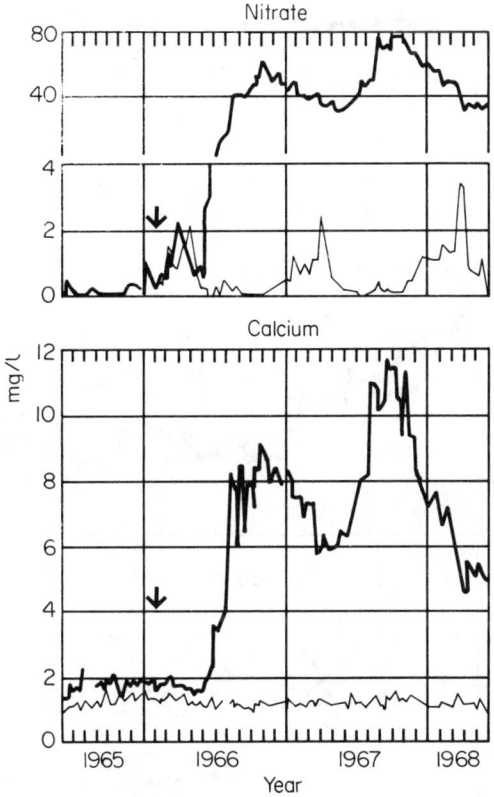

Fig. 13.3 Change in nitrate and calcium concentration in water flowing from a water shed cleared of vegetation (thick line) and an adjacent watershed with preserved vegetation (thin line). The arrow marks the date of vegetation clearance. Nitrate and calcium increased, with a 5 months delay, indicating piston flow recharge (Borman and Likens, 1970).

Monitoring of nitrate and chloride in an irrigated farm land

Saffinga and Keeney (1977) monitored Cl and NO_3 in shallow wells near a farm in central Wisconsin, USA. The results of a well in an area not irrigated by the farm (well 7) are compared in Fig. 13.4 with the results from wells located in the irrigated farm land (wells 8 and 9, the latter being situated in a less cultivated part). The farming intensity has dropped since 1972.

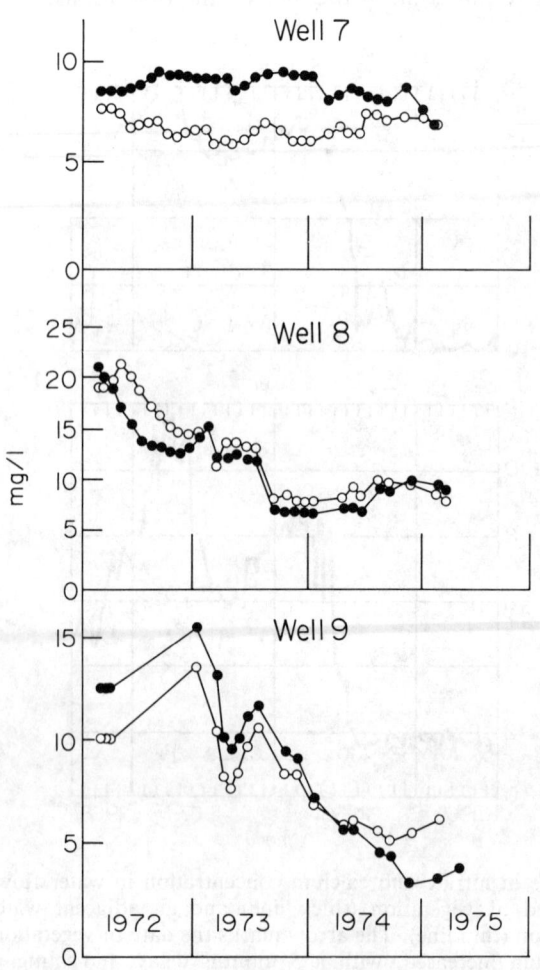

Fig. 13.4 Nitrate (•) and cloride (o) in groundwater of a well (No.7) in a non-irrigated area and wells (No.8 and 9) located in an irrigated area, Hancock Experimental Farm. Irrigation caused increase in nitrate and chloride, caused by nitrogen and potassim fertilizers, the latter containing Cl (KCl). The quality of water was resotred as a result of a decrease in farming activities (from Saffinga and Keeney, 1977).

Discussion

- The NO_3 and Cl concentrations were correlated to irrigation intensity, as no increase was observed in well 7, located in a non-irrigated area, and as seen by the drop of Cl and NO_3 concentrations with decrease of irrigation between 1972 and 1975.
- Cl and NO_3 co-varied, indicating a common source, which was found to be two fertilizers, applied in constant proportions: a nitrogen-rich fertilizer and a potassium chloride fertilizer.
- The recovery of the water quality in 1975 indicated that no salts were stored in the ground.
- The rapid recovery also indicated that the irrigation water reached the local groundwater rapidly, changes being observed from month to month (Fig. 13.4).

Pesticide contamination

Lewallen (1971) studied a farmer's well in Florida, USA. High DDT, DDE, and Toxaphene concentrations were noticed shortly after well completion. The detected source was backfill soil, brought from a location at which pesticide materials were dumped. The soil of the backfill, and sediment accumulated at the bottom of the poorly cased well, contained high pesticide concentrations. At the second half of 1967 the well was cleaned and proper casing was installed, causing an immediate drop in pesticide concentration (Fig. 13.5).

Discussion

- The contamination decreased significantly after cleaning and casing of the well, although no change was introduced to the adjacent backfill. Hence, the pesticides were fixed on the soil.
- No conduit-controlled flow of water occurred between the backfill site and the well area. Possible flow through a granular medium retarded pesticide contaminants by fixation to the soil.
- Although the quality of the water in the particular well was restored, much potential danger remained in the region and the surface and groundwater region required careful management and monitoring.

Case study of a large-scale industrial fluid waste injection

Goolsby (1971) described a thorough monitoring project of a large-scale industrial waste injection project in Florida, USA. The task was to inject fluid waste at a rate of up to 500 m^3/h and about 3×10^6 m^3/y. The fluid composition was described as an aqueous solution of organic acids, nitric acid amines, alcohols, ketones and inorganic salts.

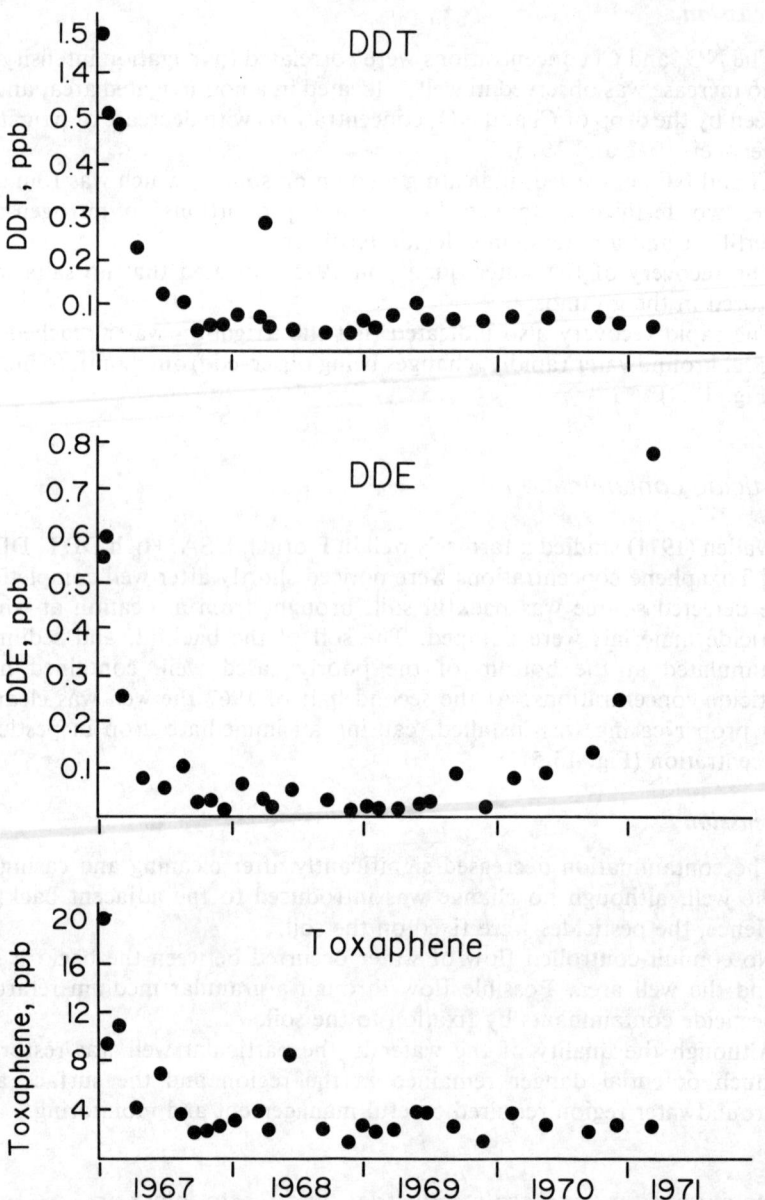

Fig. 13.5 Pesticide compounds in a well, Florida, USA. Contamination was caused by pesticide-contaminated soil in a backfill and in sediment washed into the well. Cleaning of the well, and repair of the casing, removed the contamination and the well recovered.

Discussion. The injection site was selected on grounds of the following geological and hydrological considerations (Fig. 13.6):

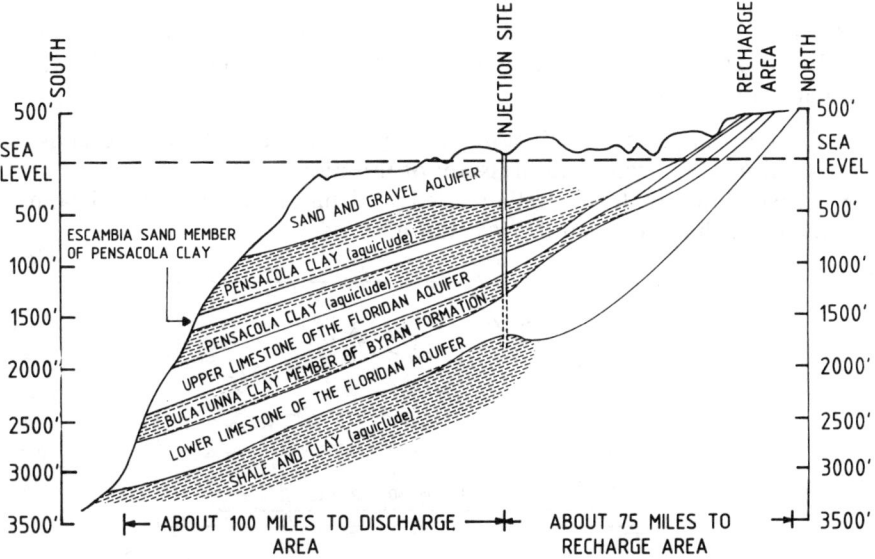

Fig. 13.6 Generalized geological cross-section through southern Alabama and north-western Florida (after Goolsby, 1971). An injection well penetrated the confined Lower Limestone aquifer.

- A free surface sand and gravel aquifer was found to overlie two confied aquifers of Floridian Limestone, the three aquifers being separated by distinct clay layers.
- The water in the upper limestone aquifer contained 425 mg/l of Cl, whereas the lower limestone aquifer contained 7900 mg/l of Cl. Thus, the two aquifers did not communicate hydraulically.
- The water of the lower limestone aquifer was non-potable.
- In no place was the saline lower aquifer water seen to ascend to the surface.
- The recharge area of the lower aquifer was about 120 km northward.
- The nearest possible discharge area of the lower aquifer, the ocean, was about 160 km south of the selected injection area.

Thus, the injection site was remote from recharge and discharge points, and the injected aquifer was saline and isolated from the overlying aquifer.

Injection and monitoring systems. An injection well A, 500 m deep, was drilled into the lower confined limestone aquifer, and later on a second well, B, was drilled 400 m west of the first one. A deep monitoring well, 400 m south of A, was drilled into the receiving aquifer. A shallow monitoring well was drilled into the upper limestone aquifer, 30 m from injection well A. Hence, changes in the chemical composition could be monitored both in the receiving and overlying aquifers.

Results. Water levels recorded in the two injection and two observation wells are given in Fig. 13.7, along with the injection rates. The following observations were made:

- The deep monitoring well responded by an increase of water level as a function of the rate (and pressure) of injection.
- The shallow monitoring well revealed no change in water level, indicating the upper limestone aquifer was not affected.

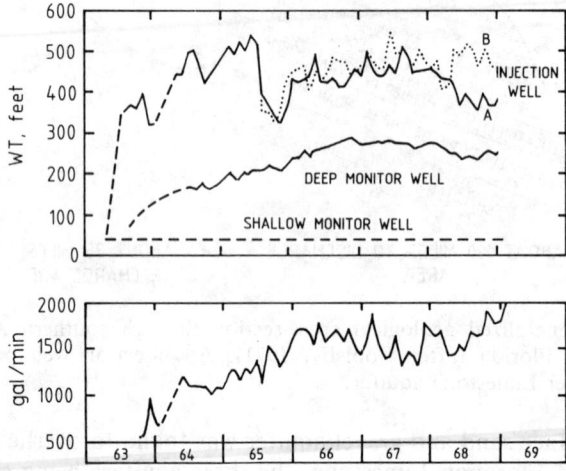

Fig. 13.7 Piezometric water levels in waste injection and monitoring wells and injection rates in Florida (from Goolsby, 1971).

The calcium concentration in the deep monitoring well is given in Fig. 13.8. The following patterns were observed:

- Although the waste fluid was low in Ca (about 3 ppm), it became calcium-tagged in the aquifer. This was actually expected from the reaction of the acid waste (pH 5, and later on pH 3) with the aquifer limestone.
- The first front of injected waste arrived 10 months after commencement of the operation. The arrival of the fluid in the 400 m distant monitoring well was noticed by a marked increase in Ca (Fig. 13.8).
- At the beginning of 1966 the Ca concentration decreased in the deep monitoring well, a feature not well understood.
- In 1968 a previous neutralizing process was stopped, and the waste was injected with a pH 3, resulting in a sharp increase in dissolved Ca in the monitoring well (Fig. 13.8). This indicated that limestone was dissolved at

a high rate. The resulting 'karstification' probably enhanced the movement of the waste in the aquifer.

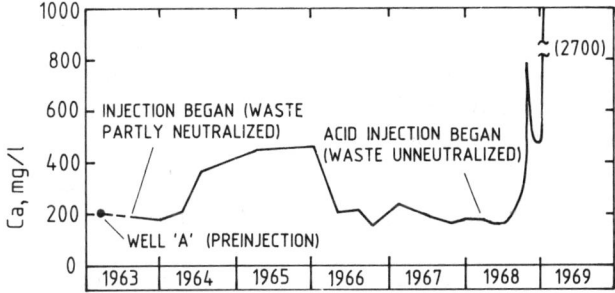

Fig. 13.8 Calcium concentration, deep monitoring well, between 1963 and 1964, Waste Injection Project, Florida (from Goolsby, 1971).

Additional chemical data are given in Table 13.1. As mentioned, the composition of injected waste was changed in the middle of 1968, when neutralization was dropped. As a result, the pH went down to 3.3 and the nitrate concentration went up. The deep monitoring well responded in the following way (Table 13.1): the October 1967 and May 1968 analyses showed high Na and Cl concentrations, revealing dominancy of the original saline aquifer water.

Table 13.1 Composition (mg/l) of waste and of groundwater (Goolsby, 1971)

	Injected wastes		Deep monitor well		
	Nov. 1967	Apr.–Dec. 1968 (Average)	Oct. 1967	May 1968	Jan. 1969
Ca	3.4	<20	168	165	2350
No$_3$	2420	5070	2	0	5760
Na	610	<1000	3720	3720	635
Cl	82	200	5700	5900	161
COD	–	22980	–	2355	22100
pH	5.2	5.3	6.80	7.00	4.75

The percentage of waste water was calculated by the investigators from the Cl concentration in the monitoring well (5800 mg/l) and the initial Cl concentration in the lower limestone aquifer (7900 mg/l):

$$\text{percentage injected water:} \frac{7900-5800}{7900} \times 100 = 26\%$$

In other words, up to May 1968 the Cl-rich aquifer water was 'diluted' by 26% of injected Cl-devoid water. The newer, non-neutralized waste, with pH 3.3, caused a remarkable change in the composition of the fluid at the deep monitoring well: Cl dropped to 160 mg/l, indicating that the deep monitoring well was dominated by 98% injected water:

$$\frac{7900-160}{7900} \times 100 = 98\%$$

The rapid increase from 26% to 98% waste fluid, in the deep monitoring well, indicated that the non-neutralized waste dissolved the aquifer carbonates and developed a highly conductive karstic system, very alarming in terms of waste disposal in the aquifer. During all this time no chemicals arrived in the upper monitoring well.

Short-circuiting of aquifers by an abandoned well

Jorgensen (1968) investigated a case of water quality deterioration in a municipal well, A, at Avon, South Dakota, USA. The city wells pumped the free water table aquifer of the Codell sandstone (Fig. 13.9). Well A had a significantly inferior quality. Pumping tests supplied the first clue: during a pumping test a contribution of 48% from another aquifer was detected. Local intrusion of water from the lower confined Dakota sandstone aquifer (Fig. 13.9) was suspected, as its water was known to be saline and of

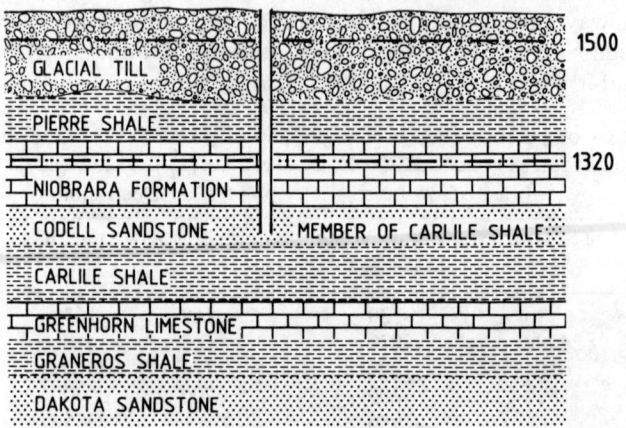

Fig. 13.9 Geologic cross-section, Avon, South Dakota, USA (from Jorgensen, 1968). An old corroded well short-circuited the deep pressurized Codell aquifer and the shallow Dakota aquifer and contaminated the latter.

piezometric head that exceeded the water level in the phreatic Codell sandstone aquifer. Chemical compositions and temperatures of the non-disturbed phreatic aquifer, the confined aquifer, and well A are given in Table 13.2. A glance at the table reveals that the properties of the pumping test sample of the deteriorated well are indeed between the properties of the water in the Dakota sandstone aquifer and the Codell sandstone aquifer. The percentage, X, of saline and warm Dakota sandstone water in the pumping test sample can be calculated (section 6.7), applying several parameters:

Table 13.2 Dissolved ions (meq/l) and temperature in wells near Avon, South Dakota, USA (Jorgensen, 1968)

	Dakota sandstone water	Codell sandstone water	Pumping test well A water	% Dakota water in A
Temperature	18°C	11°C	14°C	43
Ca	15.2	7.0	11.0	48
HCO_3	2.7	4.9	3.6	57
Mg	4.7	3.5	2.3	
Na	4.0	19.6	18.7	
K	0.51	1.04	1.05	
CO_3	0.0	0.0	0.0	
SO_4	16.8	21.6	25.2	
Cl	2.6	3.4	2.8	

by temperature: $14 = 18X + 11(1 - X)$
$X = 0.43$, or 43%
by calcium: $11 = 15.2 + 7(1 - X)$
$X = 0.48$, or 48%
by bicarbonate: $3.64 = 2.69X + 4.91(1 - X)$
$X = 0.57$, or 57%

These calculated mixing ratios agree well with the value of 48% contribution from the deep aquifer, determined from the pumping test. A subsequent inquiry revealed that in 1870 a city well was constructed only 6 m away from the deteriorated well. That well reached the confined Dakota sandstone aquifer and water rose in it above the piezometric level of the phreatic aquifer. The old well had been capped 50 years before. It thus seemed that the casing of the old well had failed, short-circuiting the two aquifers, creating a saline plume in the upper aquifer, and locally causing deterioration of water quality. The agreement between mixing percentages calculated by the pumping test, temperature, Ca and HCO_3, supported the hydraulic short-circuiting hypothesis.

Discussion. What should be done? The old well should be plugged in order to restore the separation between the two aquifers. Following the discussed mode of calculation of percentage of the saline water in well A, the data in Table 13.2 may be used to calculate the percentage of saline water in A, based on the other parameters: Mg, K, CO_3, SO_4 and Cl. It turns out that these parameters do not play the game: the mixing percentages deduced from Na, K, SO_4 and Cl differ from each other and from the values obtained by the pumping test, temperature, Ca and HCO_3.

What could cause this disagreement? Is the whole short-circuiting hypothesis erroneous? The answer is that because of the mixing of the two such different waters it is possible that secondary interactions took place affecting some of the dissolved ions. An example of such secondary reactions in disposed abattoir effluent is discussed at the end of section 13.3.

Contamination by released oil brines

Fryberger (1975) described a case study of groundwater contamination by displaced oil production brines. In 1967 a farmer lost his rice crop, near Garland City, south-western Arkansas, USA, because his well water became saline. An adjacent oil field came into production in 1955, with an increasing amount of separated brine that had been diverted into a poorly lined evaporation pond, which leaked into the local aquifer. The brine was subsequently recharged into the ground through an abandoned production well, but soon it turned out that the casing was corroded and the brine again reached the local aquifer. Eventually a new recharge well was constructed.

Monitoring project. To monitor the brine contamination, 36 observation wells were drilled to different depths. Cl, regarded as best representing the brine, was measured and the data served to produce a Cl concentration contour map (Fig. 13.10). The spread of the lines indicates that the water in

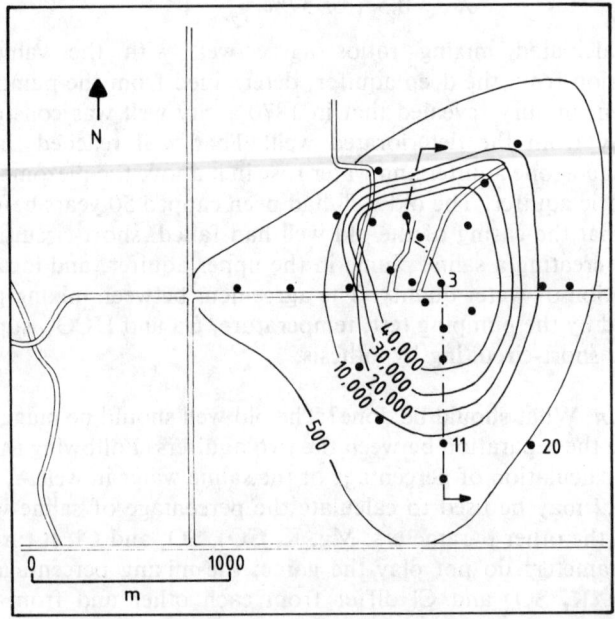

Fig. 13.10 Iso-chloride contours at the bottom of the alluvium aquifer, Gerald City (from Fryberger, 1975). Flow to the south is seen.

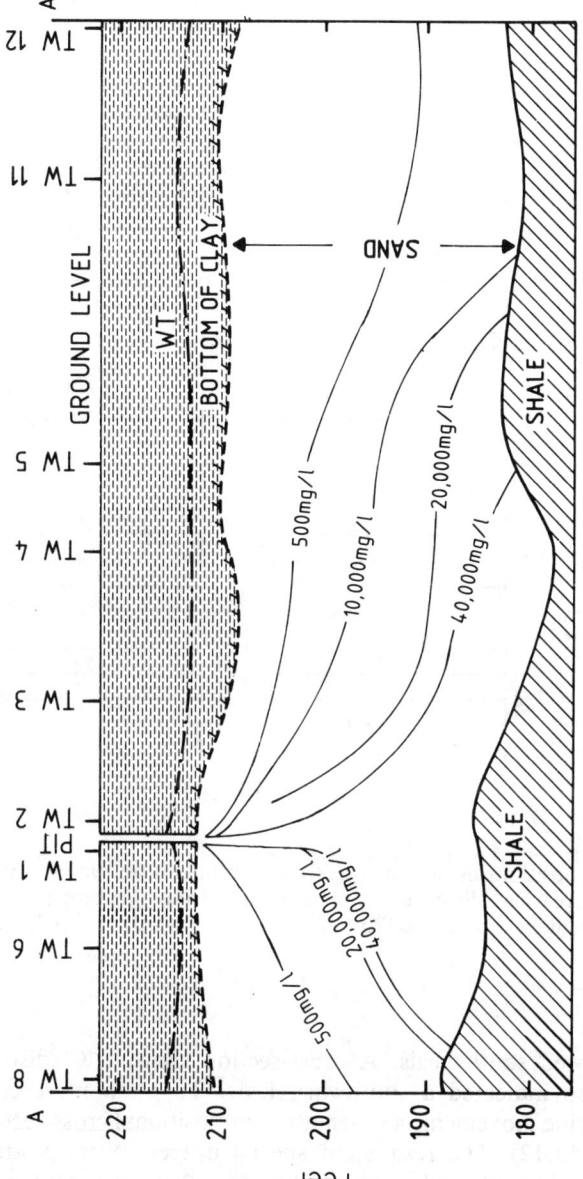

Fig. 13.11 A chloride concentration cross-section, Gerald City (from Fryberger, 1975). Leakage from the brine evaporation pit is recognizable. The heavy brine accumulated above the shale aquiclude.

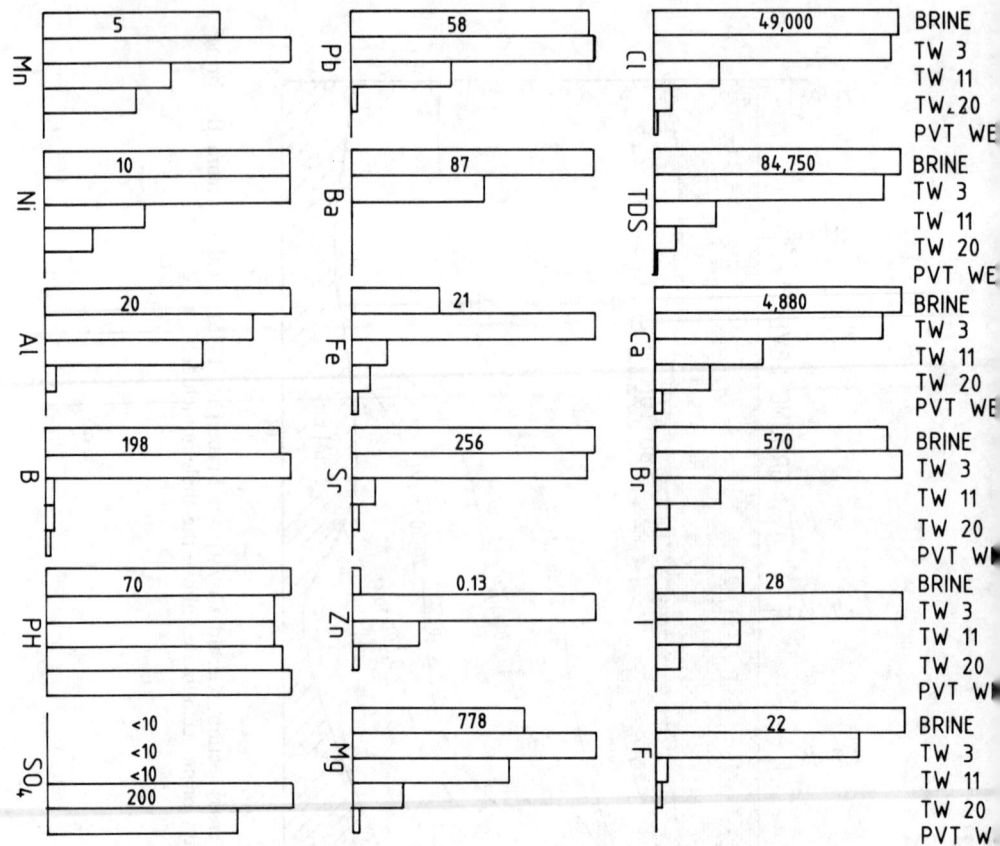

Fig. 13.12 Composition histograms (mg/l) in a brine pit and monitoring wells, arranged by increasing distance from the pit (from Fryberger, 1975). A contamination gradient is observed from the brine pit towards the fresh-water well PVT.

the aquifer flowed southwards. A cross-section showing Cl distribution (Fig. 13.11) also indicated a southward flow. To get a more complete picture of the brine movement, a schematic compositional cross-section was prepared (Fig. 13.12). The results, of special interest, were plotted as a function of distance of each observation well from the contamination centre. The concentration of Cl, Br, F, Pb, Ca, Mn, Ni and Al increased in a pattern explainable by mixing, in various proportions, of contaminating brine and local groundwater. However, several dissolved ions revealed different patterns, indicating that secondary processes occurred:

- I, Fe and Zn increased in wells over the concentrations in the brine or local groundwater. The researchers hypothesized that chemical displacement of ions coating the sand grains took place. This explanation is plausible for the Fe alone. Hence, another explanation was offered – that the brine disposed in earlier stages had a different composition.
- Ba is seen in Fig. 13.12 to drop faster than Cl in wells TW 11 and 20, indicating effective precipitation en route.
- SO_4 is high in the private well and TW 20, indicating that it is high in the non-contaminated local groundwater. The observed drop in SO_4 concentration in town wells 11 and 3 is probably due to consumption by Ba precipitation.

Discussion. The term *conservative* is applied to dissolved ions that, except for mixing, take part in no other process. In contrast, *non-conservative ions* reveal changes in concentrations caused by mixing of two water types and, in addition, by water–rock interactions. Looking at Fig. 13.12, which ions are conservative and which are non-conservative? Cl, Br, Ca, Ni behaved as conservative ions. In contrast, secondary reactions changed the composition of the water by exchange reaction with rocks, e.g. Fe and possibly Zn.

Conclusion. Different ions may travel at different velocities, the conservative ions moving fastest. The conservative ions are most helpful for the understanding of groundwater flow.

Contamination by repressuring of an oil field

Wilmoth (1972) reported that a domestic well in Kanawha County, West Virginia, USA, contained up to 1967 only 32 mg/l of Cl. This excellent water quality deteriorated because of brine injected in an adjacent oil well. Within one year the Cl concentration rose to 1140 mg/l (Fig. 13.13).

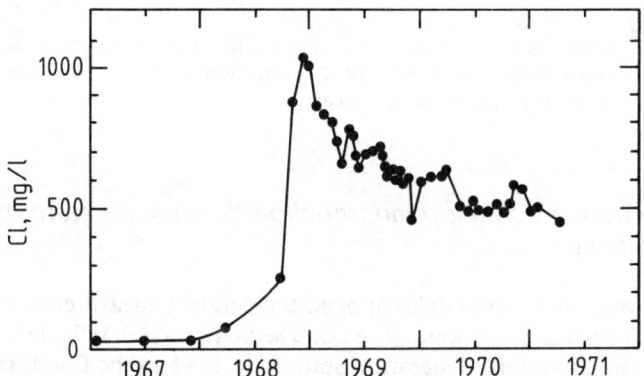

Fig. 13.13 Changes in chloride in contaminated well, adjacent to an oil field, Kanawha County (after Wilmoth, 1972). Fast response to brine injection in a nearby well was observed, as well as fast recovery after brine injection was stopped.

Operations were stopped and the Cl concentrations decreased slowly, reaching a value of 450 mg/l Cl early in 1971, so the well could be utilized once more. The decrease of Cl was caused by natural flushing by the water flowing in the aquifer.

A second case described by Wilmoth (1972) in the same county related to a fresh water aquifer near Wallace that contained 100 mg/l of Cl. In late 1967 injection of brines, as part of subsurface operations in an adjacent oil field, caused within 2 months an increase to 2950 mg/l Cl. Operations were stopped and the water quality was soon restored, attaining 190 mg/l after 2½ years (Fig. 13.14).

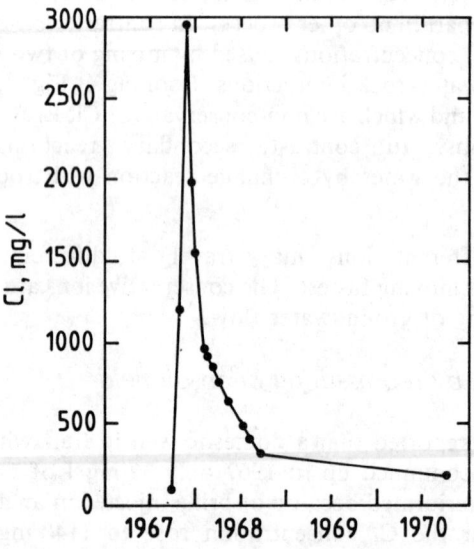

Fig. 13.14 Changes in chloride in a contaminated well, adjacent to an oil field, Kanawha County (after Wilmoth, 1972). Injection of brines into nearby wells caused a dramatic increase in Cl in 2 years.

Leakage from a cooling pond, monitored by temperature measurements

Andrews and Anderson (1978) applied temperature measurements to trace leakages from a cooling pond. The pond served two 500 MW electric power plants to the Columbia Generating Station, located on the floodplain of the Wisconsin River, near Portage, Wisconsin, USA. The cooling pond was constructed of 5 m high dikes, covering an area of 200 ha. The bottom, with silty sand, was partially lined with bentonite. The area is a discharge zone of marshes and wetlands. The geology of the area was studied by 100 borings, the water table was monitored by 80 small-diameter observation wells, and heat profiles were measured in 19 observation wells. Figure 13.15 provides an example of temperature data obtained in the extensive monitoring operation. The summer temperature peak reached well B (60 m from the pond) 60 days after it reached well A (3 m from the pond). Thus, the water leaking from the pond travelled 60 m in 60 days, or 1 m/d. A network of 47 boreholes of 10 m depth were equipped with thermistors for extensive monitoring.

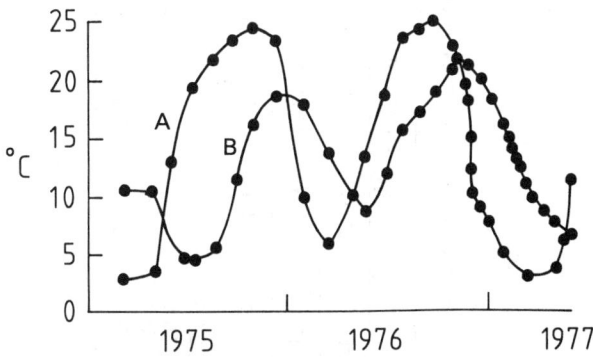

Fig. 13.15 Temperature variations in well A (4 m depth, 3 m west of dike) and well B (2.5 m deep, 60 m west of dike) (after Andrews and Anderson, 1978). The summer temperature peak reached B 60 days after it reached A, indicating a flow velocity of 1 m/d.

Contamination by a coal ash pit

Andrews and Anderson (1978) studied the impact of a coal ash pit (discussed above) on two 500 MW electric power generating units near Portage, Wisconsin. Water in the ash pit differed in its composition from the local groundwater. Operation of the power plant began in May 1975, monitored by four wells, constructed south and north of the pit dykes. Selected data are reported in Table 13.3.

Discussion. From Table 13.3, the following picture emerges: the data for 1972-3 pre-dated the operation of the power plant and reflect the composition of the non-polluted local groundwater. The variations in composition between wells A, B, C, and D reflect natural fluctuations in the local groundwater composition. For example, Ca occurred in the range of

Table 13.3 Water composition (mg/l) in ash pit and monitoring wells, near Portage (from Andrews and Anderson, 1978)

		West of the dike						North of the dike					
	Ash pit water	A – 1 m West			B – 75 m West			Ash pit water	C – 1 m North			D – 25 m North		
		1972-3	1975-6	1977	1972-3	1975-6	1977		1972-3	1975-6	1977	1972-3	1975-6	1977
K^+	2.3	1.2	9.2	6.9	1.2	0.75	0.5	2.4	1.7	0.75	1.5	2.0	0.75	0.5
Na^+	10.2	2.2	8.5	13.2	3.6	3.5	4.0	10.3	4.3	2.6	5.1	5.2	3.1	3.7
Ca^{2+}	77.8	13.9	16.9	21.6	31.6	28.8	24.3	54.7	18.8	23.3	37.0	22.0	22.6	21.5
Mg^{2+}	10.6	11.1	9.8	14.7	20.3	22.0	16.9	4.3	17.3	22.5	33.8	18.0	18.8	25.8
SO_4^{2-}	44.6	5.4	18.5	10.5	0.8	0.2	0.1	36.0	7.7	9.6	22.0	9.7	9.3	14.0
Cl^-	10.6	3.4	4.5	6.5	4.49	2.64	5.0	10.5	2.8	4.2	8.5	7.2	3.2	3.5

13.9 to 31.6 mg/l, and SO_4 varied from 0.8 mg/l (well B) to 9.7 mg/l (well D). Bearing these variations in mind, it seems that:

- Ash pit water reached no well considering K and Mg – the relatively high values observed exceeded the values in the ash pit water and must be due to some other reason (e.g. secondary water–rock interactions).
- Na indicates contamination in well A alone.
- Ca, SO_4 and Cl indicate pollution in wells A and C.

The conclusion from all observations is that the water reached wells A and C, located 1 m west and north of the pit, and no contaminating water reached wells B and D, located 75 m west and 25 m north. The contamination was restricted to the immediate surroundings only.

Looking at the composition of the water in the ash pit (Table 13.3), and bearing in mind the discussed observations on the rate of water movement, can the monitoring operations be stopped? The answer is that they cannot be stopped, and the following steps have to be undertaken:

- The water in the ash pit should be measured periodically for possible changes.
- The water in the ash pit should be analysed for a large number of possible toxic compounds, especially trace metals.
- The detailed analyses of the pit water should be used to search for more sensitive tracers.
- The conclusion, based on wells A and C, that the pit water was well confined, was based on the assumption that water moves in a granular medium with no conduit-controlled flow. This might be wrong from the beginning, or conduit flow might have evolved due to water leaking from the pit. Hence, monitoring is always needed. The composition of the water in the ash pit looked harmless from the data given in Table 13.3. It is the additional analyses recommended for this water that will determine whether it is really harmless or whether some toxic compounds are present, necessitating more careful monitoring.

Flyash (coal) contamination

Cherkauer (1980) monitored groundwater near a flyash site that operated for 8 years at the Port Washington Power Plant, south-eastern Wisconsin, USA. The maps, given in Fig. 13.16, show the potentiometric surface contours and direction of flow, TDS, SO_4, Ca and HCO_3 along with the positions of the monitoring piezometers. The ash leachate contained various metals that were not observed in the monitoring wells, except some iron.

Discussion. What can be learned from the data plotted in Fig. 13.16? The leachate monitored by the piezometers located in the pit revealed that a threat to local groundwater exists, as revealed by the concentrations of the

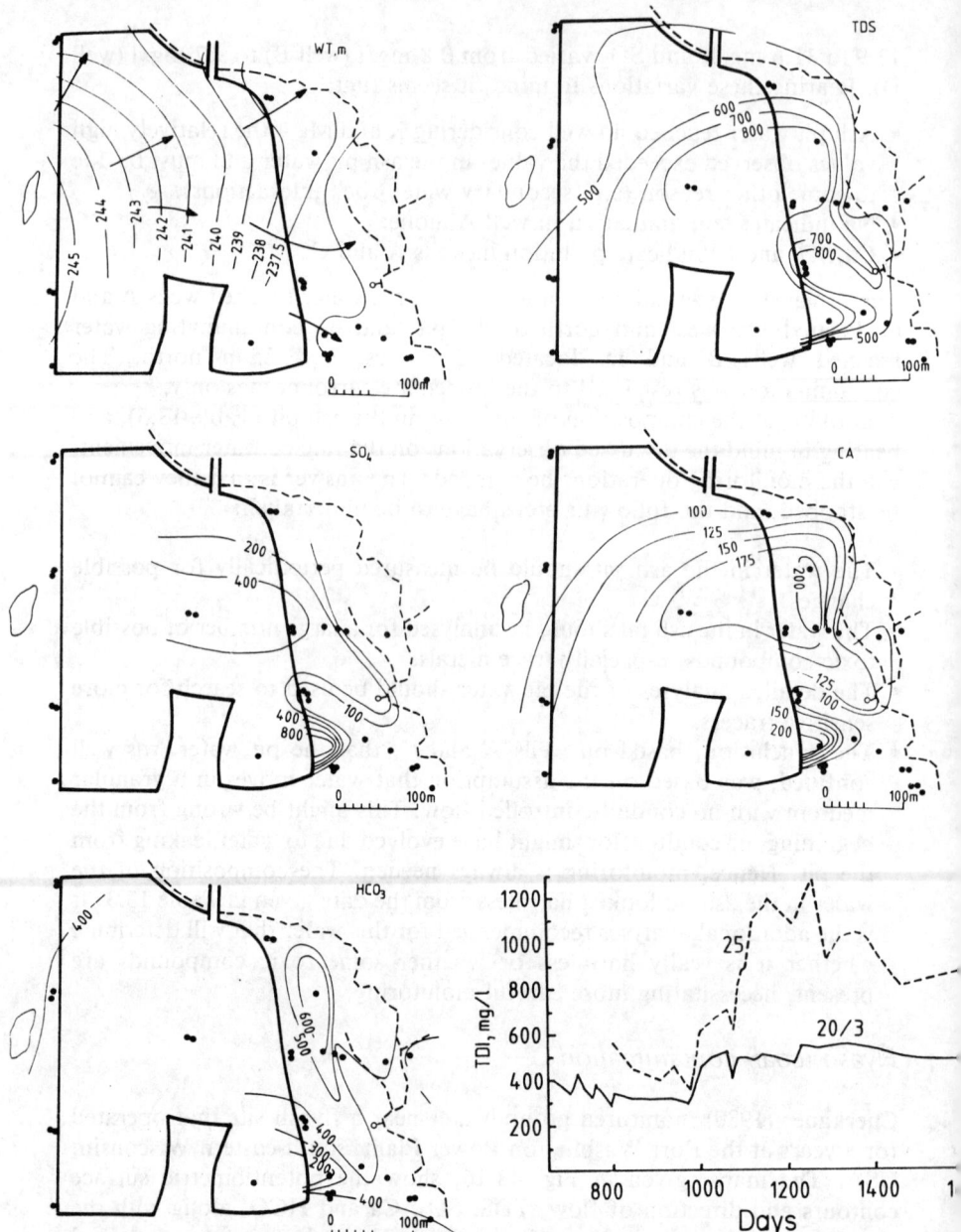

Fig. 13.16 Data of a flyash site monitoring operation, south-eastern Wisconsin (after Cherkauer, 1980). Water table contours (upper left diagram), September 1978, masl; dots mark locations of piezometers; heavy line marks the boundary of the filled area; dashed line, surface drainage. Contours of dissolved ions are in mg/l. Right bottom: TDI changes in contaminated well (25) and non-contaminated control well (20/3).

various ions. A plume of leachate entering the local groundwater is observed, in the direction of the water level gradient. Contamination is noticeable at a distance that can be estimated from the map. The observation that metals found in the leachate do not show up in the monitoring wells indicates that they are retarded by the system, most probably by adsorption onto clay material in the soil. This means that a halo of metals is formed in the soil and this halo may reach local groundwater in the future.

Retardation of poisonous metals should be taken as a temporary solution. The adsorption capacity of soil is limited, and eventually the excess metals will appear in the exploited water sources. At this stage the process may be non-reversible, i.e. stopping the recharge of the pollutants will not result in an immediate cure of the water, as the metal front will continue to come down for a long time. Hence, pollution of the aquifer by non-degradable contaminants should be stopped as early as possible. In the meantime, monitoring of wells in the area should be continued.

Monitoring leachate of a sanitary landfill

Fritz *et al.* (1976) studied possible movement of leachate from a sanitary landfill in Frankfurt am Main, Federal Republic of Germany. In a preliminary search for suitable tracers it was found that the leachate leaving the waste disposal site was enriched in D and ^{18}O (probably due to evaporation). The leachate was enriched by 1.5-2.0‰ $\delta^{18}O$, whereas the analytical resolution was ± 0.1‰. The site operated between 1925 and 1968, and about 1.8×10^7 m^3 garbage have been deposited there. The underlying phreatic aquifer, built of Quaternary sand and gravel, was biologically polluted. Samples were collected in test wells that penetrated the aquifer, at depths of 8-11 m. Samples were also collected from drainage ditches in the landfill. Partial chemical data are given in Table 13.4, along with COD (chemical oxygen demand), reflecting the concentration of organic matter in the polluted fluid. Site of test wells and piezometric controus are given in Figs. 13.17 and 13.18.

Discussion. Looking at the water table elevations (Fig. 13.18) and the analytical results (Table 13.4), which wells may be regarded as representing non-contaminated groundwater? Well 401 is upstream (Fig. 13.18) and on this ground alone should be non-contaminated. This is confirmed by the observation that the water in this well is lowest in Cl, NH_4 COD and TDS. Wells 404, 405, 406, 415, 416 and 417 are closest to the landfill, in the groundwater flow direction (Fig. 13.18). They are, therefore, to be suspected for contamination. The data reveal that these wells, and additional wells that are located down-flow, are indeed contaminated. Of special interest (and concern) is the observation that different parameters assign a different degree of contamination to individual wells. For example,

Table 13.4 Composition of surface and groundwater near landfill site, Frankfurt am Main (Fritz *et al.*, 1976)

Station No.	Cl⁻ (ppm)	SO_4^{2-} (ppm)	NH_4^+ (ppm)	COD (ppm)	TDS (by evaporation) (ppm)	Comments
401	19.3	122.5	0.16	10	256	Groundwater
404	2682	867.3	244	1310	9250	Groundwater
405	1806	89.5	1404	1070	8670	Groundwater
406	1725	990.5	603	1310	7800	Groundwater
407	987.7	285.8	623	460	6700	Groundwater
408	1670	117.1	198	880	5700	Draws lake water
409	26.0	48.1	0.16	8	233	Groundwater
410	73.1	131.3	1.55	4	493	Groundwater
411	168.9	224.7	0.74	46	722	Groundwater
412	43.1	82.4	1.78	29	344	Groundwater
413	14.6	45.6	0.06	8	348	Groundwater
414	11.7	40.5	3.09	24	287	Groundwater
415	1883	137.2	119	960	6950	Groundwater
416	1610	313.4	114	810	6300	Groundwater
417	719.3	217.1	3.19	305	2665	Groundwater
418	40.4	35.6	0.04	25	551	Groundwater
20	1636	214.3	104	820	5400	Lake water
21	1713	211.5	104	765	5400	Lake water
26	61.2	126.1	73.1	375	2011	Lake fed by ditch
28	920.5	127.9	135	520	3350	Ditch fed by groundwater

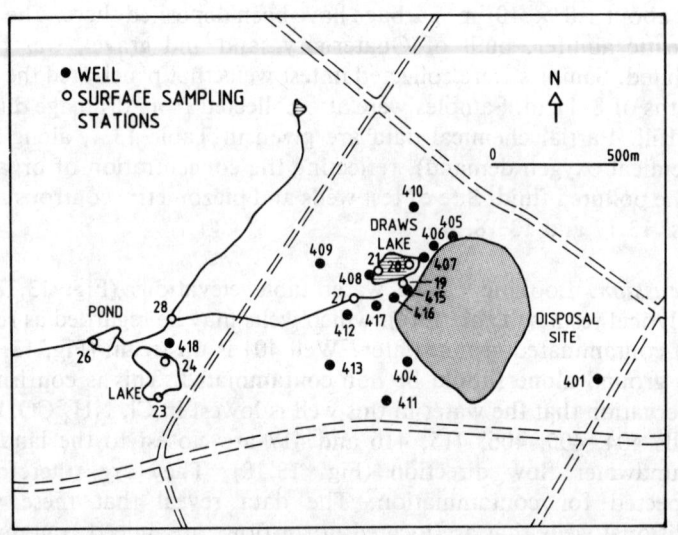

Fig. 13.17 Sanitary landfill, Frankfurt am Main (from Fritz *et al.*, 1976).

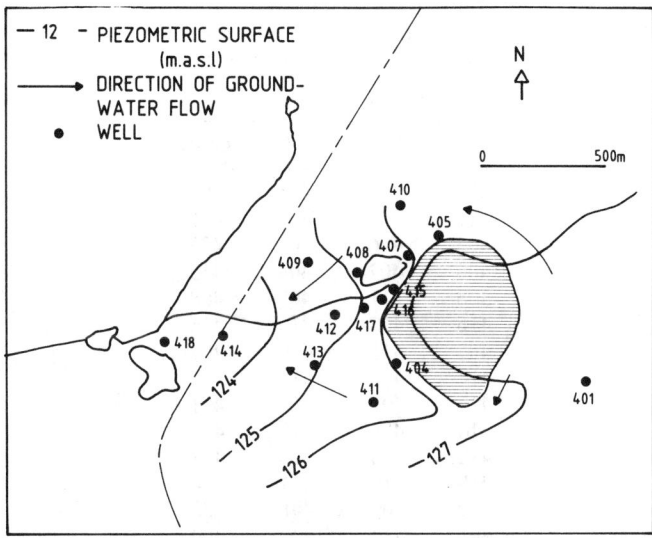

Fig. 13.18 Water levels (masl), shallow aquifer at sanitary landfill, Frankfurt am Main (Fritz *et al.*, 1976).

well 404 is most contaminated by Cl and COD but slightly less by NH_4. To make this point clear, the wells have been listed in Table 13.5 in decreasing order of concentration of each contaminant. The relative order of the wells in the Cl column is seen to differ from the arrangements by SO_4, NH_4 or COD. Yet, an overall picture is obtained: wells 404, 405, 406, 407, 408 for example, are clearly contaminated. Several reasons may cause the observed differences:

- The sanitary landfill is not homogenous, and different parts seem to contain different types of garbage.
- The non-contaminated groundwaters might have had slight local differences of composition.
- Some of the ions could be retarded by soil components, or exhanged with soil and rock, and these changes might differ locally.

Where was the contaminating groundwater front? It was situated between the most distant wells that revealed contamination and the closest wells that revealed no contamination. Well 401 has been selected as representing the non-contaminated groundwater, as discussed above. Well 401 has been emphasised in Table 13.5, and the wells listed below 401 may be regarded as non-contaminated. It is seen that by this approach only two wells are non-contaminated, taking Cl as a tracer, whereas seven wells are non-contaminated, using SO_4 as a tracer. Cl and COD were the most conservative (and therefore most sensitive) tracers in the Frankfurt am Main study. This demonstrates the importance of analysing many parameters. Later on,

Table 13.5 Monitoring wells arranged in descending order by the various tracers (after Table 13.3)

Cl$^-$	SO$_4^{2-}$	NH$_4^+$	COD
404	406	405	404
415	404	407	406
405	416	406	405
406	407	404	415
408	411	408	408
416	417	415	416
407	415	416	407
417	410	417	417
411	**401**	414	411
410	408	412	412
412	405	410	418
418	412	411	414
409	409	**401**	**401**
401	413	409	409
413	414	413	413
414	418	418	410

measurements may be limited to the established conservative, most sensitive, tracers.

Figure 13.17 shows that the landfill is situated near a pond and two lakes, which have possible influence on local groundwater, to be understood and taken into account in the interpretation of the pollution monitoring results. δD and δ^{18}O turned ot to be most useful in this respect. The δ^{18}O data for three sampling dates, given in Table 13.6, reveal three groups of values:

$-8.6‰$ to $-8.0‰$, $-6.7‰$ to $-6.3‰$, $-3.0‰$ to $-2.0‰$

Can sense be made of these three groups, denoted by δ^{18}O (and δD)? Consulting also the map given in Fig. 13.17, the following picture emerges:

- Group A, which includes well 401, is of non-contaminated groundwater.
- Group B is of polluted groundwater.
- Group C, including well 408, reflects recharge from the Draws Lake.

A second pattern, noticeable in the isotopic data of Table 13.6, is a systematic change to more negative values, seen in those wells that have been shown by the dissolved ions to be contaminated (Table 13.5). This is an indirect indication of the progressive increase of water from the landfill. The δ^{18}O data have been used by the researchers to calculate the percentage of leachate water in each well, summed up in a map of the pollution plume (Fig. 13.19). The correlation between the isotopic data and dissolved ions has been checked in Fig. 13.20. Good positive correlations are observed between Cl, COD and δ^{18}O, and almost no correlation is seen with NH$_4$ and SO$_4$. This supports the conclusion that Cl and DOD (as well as δ^{18}O and δD) were conservative tracers.

Table 13.6 Isotopic composition of surface and groundwater near the landfill site, Frankfurt am Main (Fritz et al., 1976)

Sampling point	18_O(‰) Nov 1972	18_O(‰) April 1971	18_O(‰) June 1975	δD(‰) April 1973	δD(‰) June 1975	Comments
401	−8.6	−8.2	—	—	—	Groundwater
409	−8.3	−8.3	−8.5	—	−58	Groundwater
410	−8.5	−8.6	−8.4	−55	−57	Groundwater
411	−8.2	−8.6	−8.6	—	−59	Groundwater
412	−8.3	−8.4	−8.5	—	−59	Groundwater
413	−8.2	−8.5	−8.6	−55	−58	Groundwater
414	−8.1	−8.3	−8.5	—	−58	Groundwater
417	−8.0	−7.8	8.0	43	55	Groundwater
418	−8.4	—	−8.4	—	−58	Groundwater
404	−6.3	−6.5	−6.2	—	−42	Groundwater
405	−6.7	−6.8	−7.3	−46	−45	Groundwater
406	−6.4	−7.1	−7.4	—	−47	Groundwater
407	−6.5	−7.0	−6.9	—	−43	Groundwater
415	−6.4	−6.6	−6.9	−32	−39	Groundwater
416	−6.4	−6.7	−7.3	—	−48	Groundwater
408	−2.9	−4.0	−4.4	−31	−33	Draws lake water
20	−2.0	−3.5	—	—	—	Lake water
21	−3.1	−3.7	−4.5	−24	−37	Lake water
21	−3.0	−4.0	—	—	—	Lake water
23/24	—	−3.2	−5.3	−29	−46	Lake fed by ditch
26	−7.2	−7.0	−7.8	−41	−55	Ditch fed by groundwater
26	−7.1	−7.2	—	—	—	

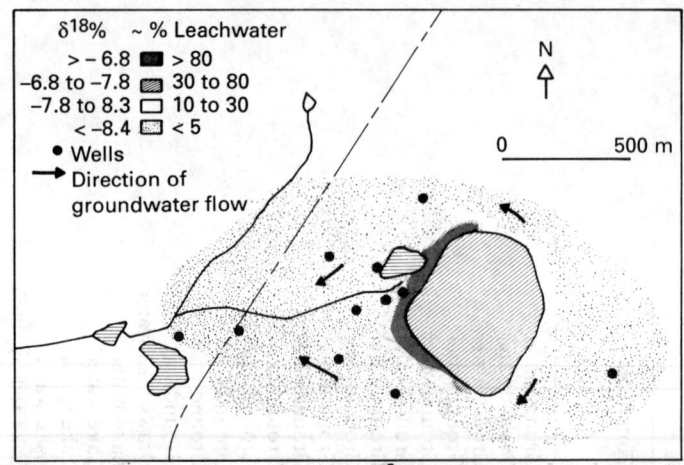

Fig. 13.19 Pollution plume, outlined by $\delta^{18}O$, landfill site, Frankfurt am Main (from Fritz *et al.*, 1976).

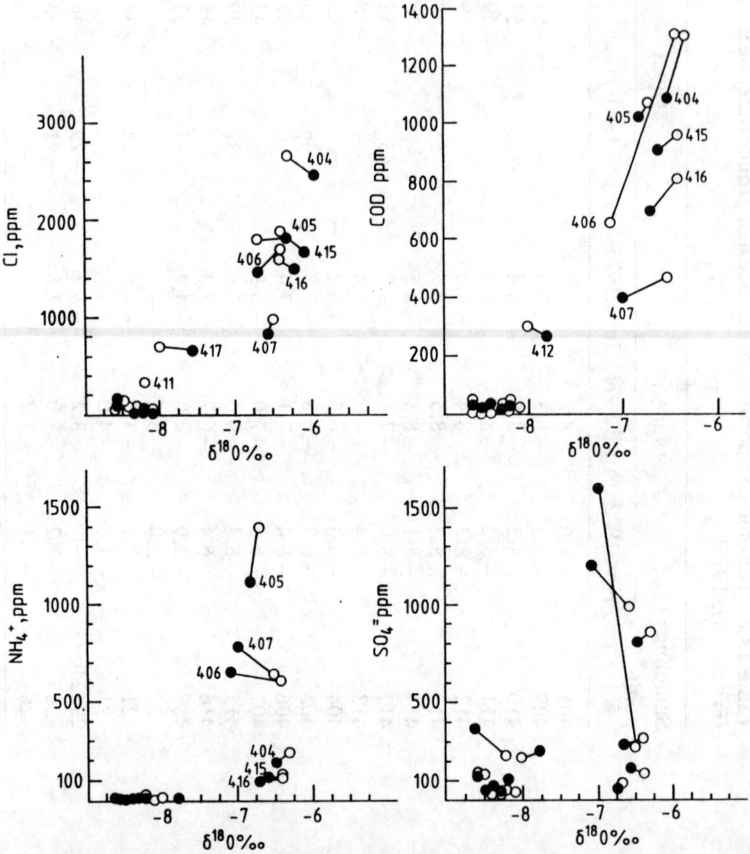

Fig. 13.20 Cl, COD, NH_4, and SO_4 as a function of $\delta^{18}O$, landfill site; Frankfurt am Main (from Fritz *et al.*, 1976). ○, November 1972; ●, April 1973.

Contamination by highway de-icing salt

Dennis (1973) studied the impact of a stockpile of 50 000 t of salt, used for highway de-icing at Mars Hill, Indianapolis, Indiana, USA. The salt, stored in 1966, was removed in 1968, because the local aquifer turned saline (several thousand mg/l of Cl). The aquifer recovered gradually. In 1973 the water of most wells was potable, but traces of the contamination were still reflected in an isochlor contour map (Fig. 13.21).

This accident supplied hydraulic information: it established the direction of flow at Mars Hill, and the observation that the wells recovered in 6 years indicates that the aquifer is highly dynamic.

A second case of contamination by road salt has been described from Monroe County, West Virginia, USA (Wilmoth, 1972). Chloride concentra-

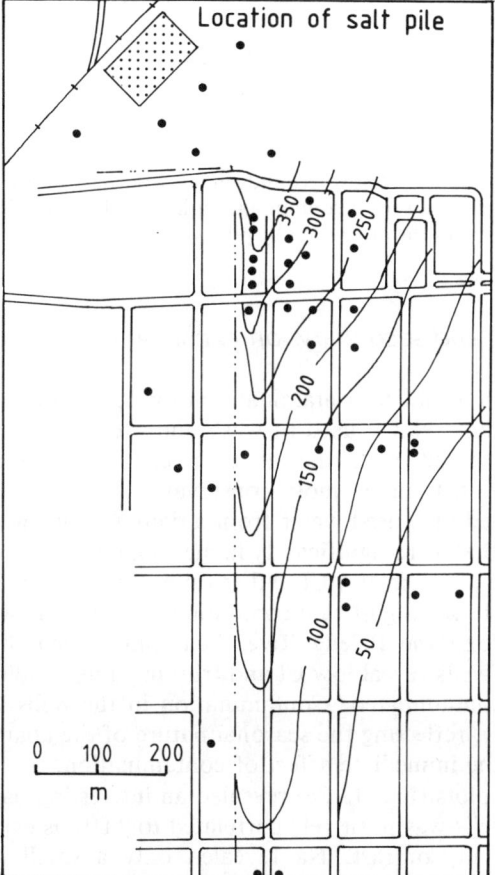

Fig. 13.21 Mars Hill, 1973 - location and isochron contours (from Dennis, 1973). Groundwater was almost recovered but the salinity distribution still reflected the southward direction of groundwater flow (see text).

tions in water of a 500 m distant well (25 m deep) increased from 185 mg/l to 1000 mg/l in 5 years. The Cl concentration rose to 7200 mg/l, in response to an increase in the salt storage. The salt piles were removed in 1970, and within 2 months the Cl concentration decreased to 188 mg/l (Fig. 13.22). The extremely fast recovery indicates that the studied aquifer has a large through flow.

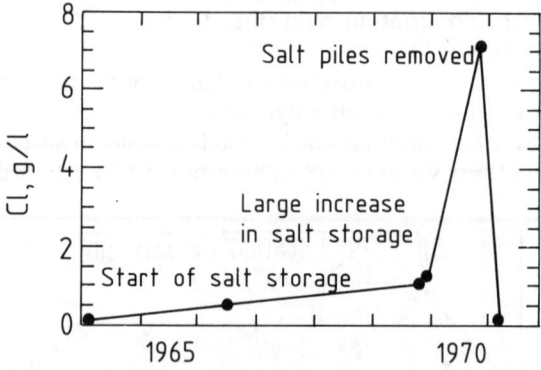

Fig. 13.22 Chloride in shallow groundwater. Monroe County (from Wilmoth, 1972). A salt pile caused severe contamination in a well 500 m distant. The system recovered within 2 months after the salt pile was removed.

Abattoir waste and secondary ion exchanges in soil

Mazor *et al.* (1981) studied contamination of groundwater by an abattoir at Lobaste, Botswana. NaCl, used to wash and preserve skins, was a major contaminant released with the abattoir effluent. Preliminary studies revealed Cl and tritium as most conservative tracers in this case. The abattoir, it turned out, used water from a dam and this water had tritium concentrations that were significantly higher than in the local groundwater. Location of the abattoir and Cl and tritium concentrations are shown in Figs. 13.23 and 13.24. Significant contamination is seen in wells close to the abattoir and along the Peleng River bed, into which the effluent was released. Other wells reveal low Cl and tritium values, indicating that they remained non-contaminated. Contamination in the wells fluctuated with time (Fig. 13.25), reflecting the seasonal nature of the abattoir output and demonstrating the immediate affect of contamination.

Composition plots (Fig. 13.26) revealed an interesting pattern: Cl in the contaminated wells was positively correlated to TDI, as expected from the NaCl disposal. In contrast, Na revealed only a small increase in the contaminated wells and no correlation with TDI. This was compensated by increase in Ca and Mg in the contaminated wells (Fig. 13.26). The explanation offered was that Na was exchanged for Ca and Mg upon

Monitoring of Contaminants 249

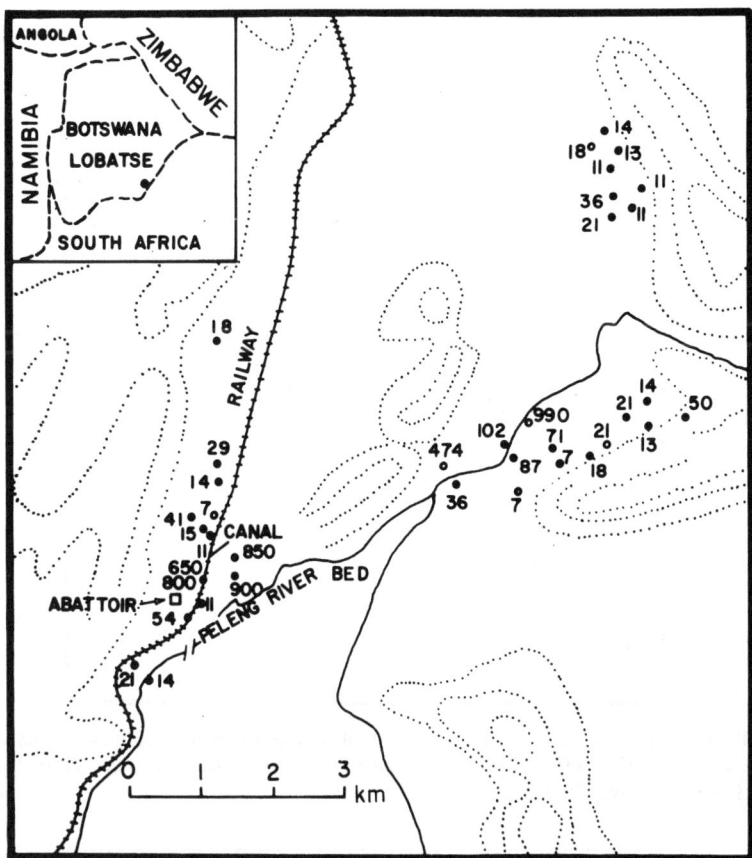

Fig. 13.23 Chloride concentrations (mg/l) observed in wells (•), Lobatse, Botswana (from Mazor *et al.*, 1981). High concentrations were observed near the abattoir and close to the Peleng river bed.

interaction of the abattoir effluent with Ca and Mg carbonates, or with clays, in the local soil.

The observation of Na exchanging for Ca and Mg is of general importance – it is, for example, observed in certain cases of sea water intrusion (Mazor, 1978). The observation at Lobatse was, thus, another case in which a pollution incident served as a long-scale tracing experiment, checking a natural phenomemon. Because of its special interest, the exchange was checked in laboratory experiments in which the sewage effluent was stored with Lobatse soil samples for 170 days (Table 13.7). The results are plotted in Fig. 13.27, revealing a drop in Na and gain in Ca and Mg. Such exchanges are common, and *a contaminant released into the ground may cause pollution of groundwater by another contaminant.*

A second example of exchange is seen for the Lobatse case in Fig. 13.26.

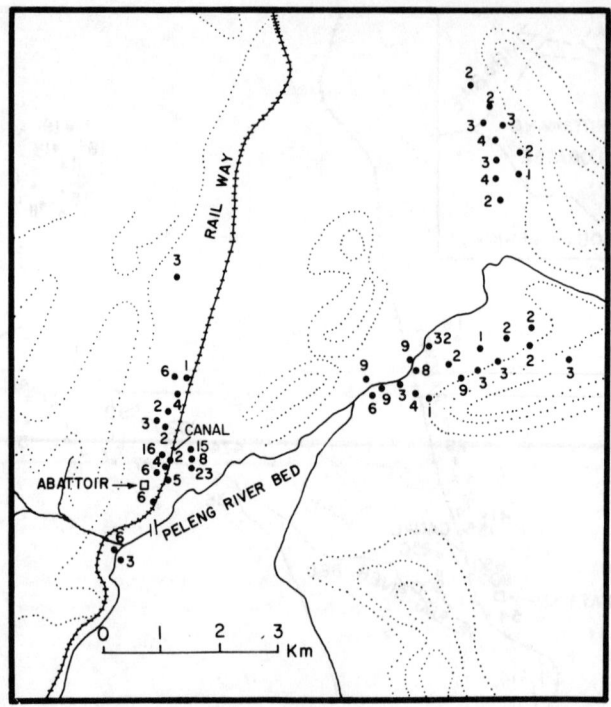

Fig. 13.24 Tritium concentrations (TU) observed in wells, Lobatse, Botswana (from Mazor *et al.*, 1981). High values were observed in wells near the abattoir and near the Peleng river bed.

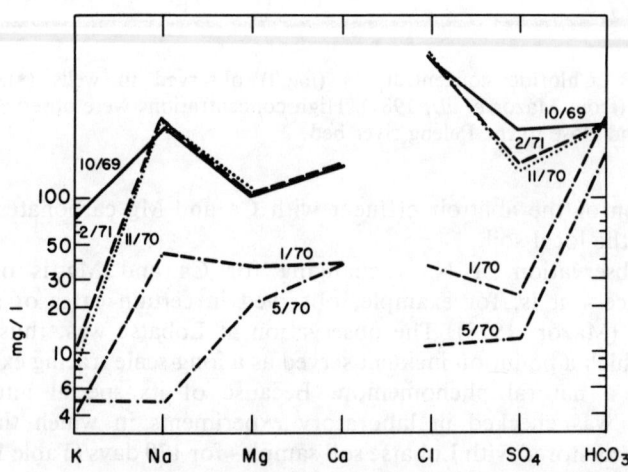

Fig. 13.25 Variations in dissolved ions in well P209, located on the abattoir grounds, at Lobatse, Botswana, reflecting the seasonal nature of the release of abattoir effluents.

Monitoring of Contaminants   251

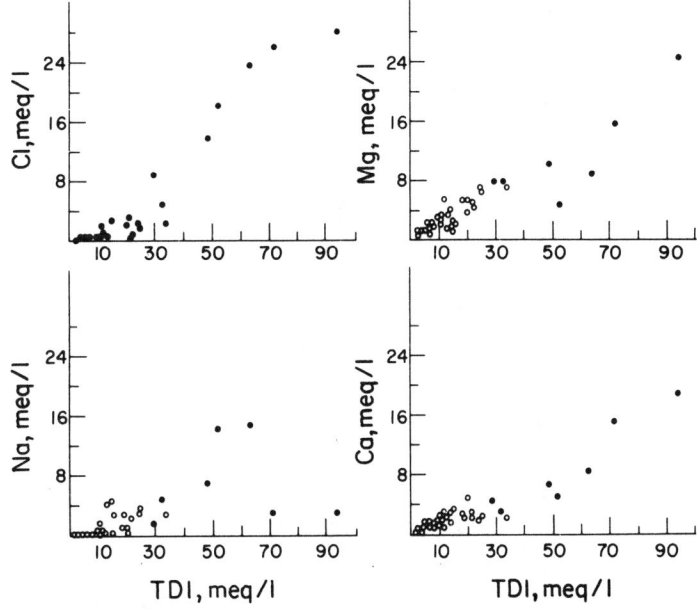

Fig. 13.26 Composition diagrams for wells at Lobatse, Botswana (from Mazor *et al.*, 1981). Non-contaminated wells (o) and contaminated wells (•). Na does not reveal an increase with TDI, but Ca and Mg do, indicating that Na was exchanged for Ca and Mg.

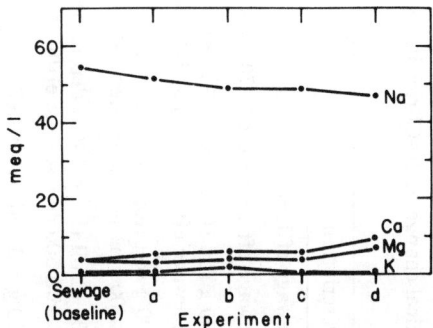

Fig. 13.27 Laboratory experiments: compositional changes in NaCl-rich industrial and domestic sewage after 170 days contact with soils of Lobatse, Botswana (Table 13.7) (from Mazor *et al.*, 1981). Sodium was exchanged for Ca and Mg.

Table 13.7 Chemical changes in NaCl-rich industrial and domestic sewage stored in the laboratory with soils of Lobatse, Botswana (Mazor et al., 1981)

Experiment	Description	K⁺	Na⁺	Mg²⁺	Ca²⁺	Cl⁻	SO₄²⁻	HCO₃⁻
Base line	Sewage effluent	1.2	52.6	4.7	3.6	59.2	4.4	4.2
	Sewage effluent stored 170 days ('blank')	1.0	54.4	4.0	3.9	60.3	5.5	2.2
a	400 g soil (from well 1931 area), with 500 cm³ sewage effluent, stored 170 days	1.1	51.1	3.4	5.5	58.1	4.3	0.33
b	400 g soil (top soil, near Geol. Survey) with 500 cm³ sewage effluent, stored 170 days	2.0	47.9	4.0	6.2	58.9	5.0	0.46
c	400 g soil (from 1.8 m depth, near Geol. Survey) with 500 cm³ sewage effluent, stored 170 days	0.51	48.9	3.9	6.2	58.3	2.9	0.24
d	400 g red soil (from well 1379) with 500 cm³ sewage effluent, stored 170 days	0.46	46.8	6.9	9.4	56.2	4.0	0.33

The SO_4 increased along with other ions. The explanation: the NaCl effluent enhanced dissolution of secondary gypsum in the soil and aquifer rocks.

13.4 Nuclear waste disposal

Disposal of radioactive waste products is an ever-growing concern for the human race. The waste of nuclear fuel processing plants, military installations, nuclear power reactors, and a large number of industries that apply radioactive isotopes, is either dumped in unsuitable locations, or is accumulated in containers, waiting for disposal in authorized repositories. The major concern is pollution of groundwater. The topic occupies large research teams in many countries, it is the subject of active international cooperation, and the results are to be found in an enormous body of literature, including special journals.

The present book, dealing with applied chemical and isotopic groundwater hydrology, addresses the basic concepts, tools, and modes of data processing, needed to deal with nuclear waste disposal. The basic principles are discussed below, in terms developed in the preceding chapters.

Several strategies have been suggested for the removal of nuclear waste of different kinds of varying degrees of concentration. These include injection in wells above phreatic water systems or deep below such systems, injection into confined aquifers containing saline (non-potable) water, burial in solidified forms in deep galleries, etc.

The mode of proposed disposal will vary from one case to another and the factors taken in account may be political, legal, economic, or technical. *The task of the hydrologist is to understand the relevant water systems and provide answers needed by the decision-makers.*

The information is needed in terms of hydraulic interconnections; nature of system – phreatic or confined; number of water types involved; mode of recharge and underground flow – through porous media or conduits; and most important – water age.

The last point warrants some discussion: all radioactive waste is planned to be buried safely, to avoid any dangerous impact even in the distant future. This is accepted as a moral obligation we have to future generations. Thus, groundwater dating is a major part of any hydrological assessment of a potential nuclear waste repository. Isolation of the waste for time scales beyond the limit of the ^{14}C method are commonly required; hence the potential importance of the ^{4}He and ^{40}Ar groundwater dating methods (Mazor and Bosch, in press) and the interest in uranium disequilibrium, ^{39}Ar, ^{81}Kr and ^{36}Cl methods (Loosli and Oeschger, 1978; Bentley and Davis, 1980). The reader is referred to Management of Low-Level Radioactive Waste (1979), and: *Isotope Techniques in the Hydrological Assessment of Potential Sites for the Disposal of High-level Radioactice Waters* (IAEA, 1983).

13.5 Summary of case studies

Detection of sources of contamination, and monitoring of waste disposal of all kinds, are tasks that have to be handled with the highest standards of expertise. No case is identical to previous cases, yet experience plays a major role. The case studies described in the present chapter are a small assortment – just to convey the nature of the job. A strong recommendation is to read and become familiar with as many case studies as possible. Those published in professional journals have the advantage of being well presented and concentrating on the highlights. Most pollution research is summed up in bulky reports that can be obtained from relevant agencies. Such reports provide a more realistic picture of the difficulties that have to be overcome, the huge amount of data accumulated and a countless number of possible modes of data processing and problem solutions.

Pollution detection and monitoring projects should be planned and conducted along the guidelines that have been found useful for all hydrochemical studies (Chapter 7). To these should be added the requirement for specialized laboratories to analyse the specific compounds present in unusual wastes.

14 HYDROCHEMISTS' REPORTS

14.1 Why reports?

Scientific research has to be summed up in a written format with appropriate circulation. The final stage of any research is a report - needed for the following reasons:

- To expose the work to critical comments by other specialists.
- To convey the findings to the authorities that financed the investigation.
- To convey the results and recommendations to the engineers and other researchers who have to execute the recommendations.
- To document the findings for future investigators who may come back to the topic.
- To share your experience with the international scientific community.

In addition to these justifications for the preparation of written reports, there is a further benefit in report writing that cannot be overestimated: the preparation of the report clears the researcher's mind, clears up internal inconsistencies and focuses thought on the essential issues. Thus, *preparing reports is an integral part of the scientific thinking process.*

14.2 Types of report

Written reports vary in length from a single page to hundreds of pages. The extent depends on the type of report and the specific nature of each study.

A research proposal is in most cases the first document to be written in any investigation. It should focus on the problem to be addressed, work done so far, strategy of the investigation, suggested schedule, organizations involved, and required budget. A research proposal has to be brief and contain only the most important points.

Progress reports are required periodically by regulations of the financing agencies. Beyond that they are of prime importance to researchers themselves. Progress reports force people to realize that time passes and data have to be processed and discussed continuously. Conclusions reached in progress reports influence the continuation of the study, as priorities are reassessed. Progress reports have to contain substance – results and preliminary discussion of data. The length of progress reports is basically defined by the amount of data and maturity of their processing reached over the reported period.

Final reports are what the name suggests – final. In other words, they should contain all relevant information and be comprehensive. Final reports are the most extensive of all reports and may sometimes contain hundreds of pages, although these are exceptional cases dealing with large-scale projects. They may have an outline similar to scientific papers but supported by all the raw data, mostly organized in appendixes. Because of their extent, final reports should contain an abstract.

Scientific papers are often an outcome of a final report. The scientific paper has to be concise, a requirement that may be achieved by referring to the final report for details. A scientific paper should deal with a local problem, but in a mode that is applicable elsewhere. Thus, scientific papers have a methodological aspect which is of much importance. In optimal cases the specific study serves as a case study to demonstrate the methodology of the scientific approach.

No recipes exist for writing reports – each case has to be solved for itself. The following sections deal with the various parts of a report in a generalized form. They are to be regarded as guidelines.

14.3 Internal structure of reports

Name or title

The name of a project is, in a way, a one-sentence summary of the most essential parts of the whole study. Thought has to be given to proper formulation of the name of the project, the report, or the scientific paper. This is best done by first listing the key words that should be included:

- The main issue: pollution, recharge or discharge estimates, mixing of water types, groundwater dating, pumpage consequences, or general survey.
- Methods used: pumping tests, chemical analyses, isotopic analyses, tritium and ^{14}C dating, trace elements, specific polluting compounds analysed, monitoring, water table data, or temperature surveys.
- The geographical location of the study.

Below is a list of sample study titles, including major key words:

- Chemical and isotopic survey of springs and wells in the Paradise Valley.
- Flow direction in the Apollo aquifer, Utopialand, traced by a landfill leachate.
- Leakage of the Fantasia Dam, traced by environmental water isotopes.
- A groundwater transect through the Episode Valley: hydrological, chemical, and isotopic parameters.
- Identification of recharge areas of the Maximum Springs, applying stable isotopes, tritium, carbon-14 and helium.
- Paper factory as possible source of salinization of the Alpha aquifer, Betaland.

Abstract

The abstract provides a brief report of the major outcome of the research. It does not deal with the history of the project, the type of measurements conducted, or hypotheses that could not be confirmed. The abstract should avoid non-informative sentences. For example, what is wrong with the following abstract:

> "An intensive study has been conducted on a large number of water sources in the study area. The paper deals with the various aspects of the main problems, leading to a series of practical recommendations".

This example contains no single item of information, and reading the abstract gives no idea of the paper. Non-informative words are: 'a large number', (instead of '21 wells'); 'water sources' (instead of: 'a lake and 12 springs); 'study area' (instead of: 'Wisdom Valley, Bestland'); 'The paper deals with' (should be omitted); 'various aspects' (instead of: 'consequences of clearing of the Green Forest'); 'leading to a series of ...' (should be omitted); 'practical recommendations' (instead of: 'limiting forest clearing to no more than 10% of the area is recommended'). The abstract should contain the most important data, e.g.:

> "The Curiosity aquifer is saline (1500 to 2200 mg/L TDI), sodium–bicarbonate dominated, with temperatures of $59 \pm 2°C$, devoid of tritium, carbon-14 concentrations are 6-8 pmc, δD is $-120 \pm 10‰$ and $\delta^{18}O$ is $-16 \pm 2‰$. Recharge from the Top Mountains is concluded, depth of circulation being >850 m, effective age 22000 ± 3000 y. So far no invasion of the Filthy Landfill leachate has been noticed, based on the chemical and biological data."

Targets of the study

Hydrochemical studies may be motivated by pure scientific curiosity, urgent management problems, or routine surveys. Behind each of these motivations stand one or several targets, and they should be spelled out. There is a saying that 'a good question is half its answer'. By the same token, it may be said that *well-defined targets promise well-conducted research focusing on the major issues*. Which of the following sentences are well-defined hydrochemical research targets?

- Checking hydraulic interconnections between the Grand Dam and the Greytown municipal well fields I and IV.
- Conducting general observations on the local hydrological system.
- Testing the applicability of carbon-14 dating techniques to the non-carbonate aquifers of the Glorious Groups in the Papaland.

The first and last examples are well-defined hydrochemical research titles.

Definition of the study area

Terms like 'the northern Great Artesian Basin', 'surroundings of London', or 'arid zones of Israel' may be suitable for the name or title of a project, but in the report or the paper itself the study area should be precisely defined. The areal definition may be by co-ordinates or by names of geographical boundary objects. The study area should always be defined on a clearly readable map, along with a key map that shows the study area in a larger geographical frame, e.g. country.

Hydrological inventory

The report should have at the beginning a section describing the hydrological inventory of the study area: sea coast, lakes, marshes, rivers, dry river beds, springs, abandoned wells, operating wells, observation boreholes. Sources of available information on these water bodies should be discussed, e.g. published papers, reports issued by various authorities, research theses submitted at universities, and data stored in archives. The exact titles should be included in the reference list of the report. To each source of information a sentence or two should be added on the nature and extent of the included information, e.g. water level measurements, discharge values, periodical quality checks, borehole and casing data, or chemical and isotopic analyses.

Human activities and installations relevant to the study

A clear and concise description should be provided on relevant human activities in the study area. These should include dams, drainage channels, recharge installations, sewage ponds, industrial effluent ponds, fluid waste disposal installations, pumped well fields, nature and extent of agricultural activity including irrigation schemes and use of fertilizers (types, quantities) and pesticides (types, quantities). Part of the information may be abstracted from detailed maps, but most information has to be obtained directly from local authorities, farmers and industry.

Strategy of the investigation

A good investigation is not carried out by shooting in all directions. It uses an outlined strategy. A clear distinction has to be made between known facts and unknowns, between established methods that have already been tested and calibrated, and developing methods that have to be checked and calibrated in hydrological systems that have been understood with other, established, methods.

Formulation of a preliminary conceptual model, based on existing data, is most helpful in defining the strategy of an investigation. Such a preliminary conceptual model should include alternative recharge areas, modes of flow, flow paths, mixing combinations, sources of heat and dissolved solids and dissolved gases, and alternative sources of potential contamination. In a second stage criteria should be defined by which wrong alternatives may be ruled out and right alternatives may be proven. At the third stage the research strategy may be formulated, focusing efforts on necessary observations and collection of required data. The following is an example of a study investigation strategy, taken from the introduction of a paper on research of CO_2-rich springs:

> "The strategy of data processing was (a) searching for water groups of common chemical composition, (b) searching for geological or geographic logics for the chemical groups established, turning them into geochemical water groups, (c) in each group the distribution pattern of ion concentration variations was studied. Mixing lines emerged and the properties of the endmembers were qualitatively defined, (d) once it was found that HCO_3 is the dominant anion, accompanied by different cations, possible water-rock interactions were explored to explain the identified geochemical water groups, (e) the stable hydrogen and oxygen data were processed to look for the role of local recharge (lying on the meteoric line) versus waters with modified isotopic compositions (e.g. $\delta^{18}O$ shifts). For water lying on the meteoric line the average recharge altitude was calculated from the isotopic composition, (f) the tritium data were interpreted in terms of pre- and post-nuclear bomb tests water components, (g) the temperature data were interpreted in terms of depth of circulation of the groundwater endmembers, (h) $\delta^{13}C$ values, obtained for the CO_2 emanating in a few mofettes, and bubbling in a few springs, were applied to define constraints on the possible origin, e.g.: decomposition of organic material, heat induced decomposition of

carbonatic rocks or mantle contribution. This choice of CO_2 origins was narrowed by noble gas data: non atmospheric He is observed in the springs, explained as accompanying the ascending CO_2. $^3He/^4He$ ratios of 0.1 of the atmospheric ratio ruled out a mantle origin, leaving both the He and CO_2 as of crustal origin."

Results

Data are a major product of every investigation. Our interpretations, conceptual models, and recommendations may all be challenged by other investigators or turn out to be wrong from our own subsequent work. Observations and measurement results may, however, always be incorporated into new models and coupled with additional observations. Hence, results are a most important part of a report or paper.

Results should be presented in a way that makes them useable by other investigators. Data are best presented in well-organized tables (section 6.1), along with relevant technical data, such as: who measured each parameter, measuring error, limit of detection, resolution, results of duplicate and triplicate measurements, measuring methods applied.

Many of the data tables may be included in the appendixes of a report.

Data processing and interpretation

Hydrochemical data may be processed in a variety of modes and accordingly presented in different ways. The latter include fingerprint diagrams (section 6.2), composition diagrams (section 6.3), histograms, maps, transects (section 7.9), and a host of other graphic modes.

Presentation of processed data is closely linked to interpretation, as the obtained patterns reflect distinct boundary conditions, e.g. number of water types occurring, mixing of water types and mixing percentages (natural and polluted systems), hydraulic interconnection and hydraulic isolation, positive and negative correlations between parameters, directional change in properties (on transects and maps), indicating directions of flow and modes of intermixing.

Much thought should be devoted to determining the best sequential order in which processed data should be presented in a report. The order should be defined by the study strategy and logics of the case presentation. In a poorly organized report, use is often made of sentences such as 'this point is discussed later on' or: 'the relevant data are presented in a later section'. In a well-organized report topics are discussed in the right progressive order. The data processing and interpretation sections of a report lead to sub-conclusions, or define constraints, all to be taken into account in the formulation of the overall conceptual model.

Exact terms, logic and informative statements

Conclusions, boundary conditions and conceptual models have to be formulated with the aid of exact and relevant terms.

Examples
- What is wrong in the following statement?
'The shallow wells contain fresh water and therefore contain post-bomb tritium.'
  The term 'fresh water' relates to *low TDI concentrations*, whereas post-bomb tritium relates to a *recent age*. The presence of post-bomb tritium *indicated* that the water is of a recent age, and not vice versa.
- What is wrong in the following sentence?
'The low temperature of the springs indicates dominance of a recent water component.'
  Low temperature may indicate dominance of a cold water end member, but *temperature* does not relate to *age*.
- What is wrong in the following sentence?
'The water is old, as indicated by its high salinity.'
  High salinity may be caused by many processes (evaporation, intermixing with saline water, dissolution from rocks) and is not an indication of a high groundwater age.
  Logic has to be maintained in every sentence of a report. *Cause* and *outcome* should not be confused; neither should *observation* and *hypothesis*.
- Correct the following sentence:
'The new deep wells tap mixed waters and therefore have discordant ages'.
  Observation (discordant ages) leads to the *conclusion* (the water in the wells is a mixture) and not vice versa.
- Correct the following statement:
'Piston flow is assumed to dominate in the Good Hope wells that have seasonally varying temperatures.'
  The *observation* (seasonally varying temperatures) leads to the *conclusion* (dominance of piston flow) and the term 'assumed' is out of place.
Statements in a report should always be informative as much as possible.

Examples
- How can the following sentence be made informative?
Group A waters are warm and group B waters are cold?
  Warm and *cold* are relative terms, but providing the relevant temperatures will make the sentence informative:
  'Group A waters are warm (52-64°C) and group B waters are cold (15 – 17°C).'

- Is the following sentence good enough to be included in a hydrochemist's report?
  'Magnesium is correlated to bicarbonate in two out of the three water types prevailing in the area.'
  No, this sentence does not qualify: 'Is correlated' - negatively or positively?; 'two out of the three groups' - so in which is the correlation obvious?; 'in the area' - which? A correct version would be:
  'Magnesium reveals a linear positive correlation in water groups I and III of the Grand Scheme project area.'

Conceptual model

Conceptual models (section 1.4) should be presented in a clear, logical, and concise way. Each part of the model should be based on a specific sub-conclusion or constraint reached in the previous parts of the report. A conceptual model should discuss alternatives and explain criteria used to select the preferred possibility.

A conceptual model should, as far as possible, be formulated in terms that will easily lead to conclusions and recommendations. Therefore, discussion of a conceptual model should also include definition of weak points and open possibilities.

The conceptual model is, in a way, the heart of every investigation and much thought should be devoted to its construction and proper presentation. A general graphic scheme is often useful.

Quantitative (or mathematical) model

Parts of a conceptual model, or all of it, may in many cases be quantified, e.g. by mass or energy balances, chemical equilibrium and degree of saturation calculations, or flow velocities calculated by hydraulic parameters, versus velocities calculated via isotopic age indicators. In presenting the mathematical calculations *knowns* and *unknowns* have to be clearly defined and *basic assumptions have to be discussed*. Each calculation should be accompanied by an evaluation of its *degree of confidence*.

Scientific conclusions

Models may be followed by a discussion of the scientific conclusions reached in the study. For example: which elements and compounds tend to behave as conservative parameters; which ions participate in secondary water - rock interactions; reliability of isotopically based calculations, e.g., recharge altitude (section 9.8) or $\delta^{13}C$ values as a tool to correct observed carbon-14 values for water-rock interactions.

Scientific conclusions are of a local nature but have a general value, because they may be applied to studies in other regions as well.

Operational recommendations

Operational recommendations may have a practical aspect, e.g. rate of suggested pumping, spacing of new wells and their required depths. Operational recommendations may also relate to further research needed, e.g. new parameters to be measured, special pumping tests, or new wells. Recommendations have to be clear, and their reasoning should be explained.

Example. Does the following recommendation fit the described specifications:

> 'A few deeper wells should be drilled to get a better understanding of the system.'

This recommendation is totally worthless: 'a few' — how many?, what are the technical specifications?, where?; 'to get better understanding' — how? by closing large gaps between existing wells? by better installation of the wells, e.g. casing opened to specific aquifers? by proper pumping tests?

Recommendations for further research should be given only if they are fully justified. Vague words such as 'deeper', 'more', or 'better' should be replaced by explicit definitions.

Recommendations should also be weighted in the light of the costs involved.

REFERENCES

Textbooks for further reading are marked by an asterisk (*), and special texts published by the International Atomic Energy Agency are given at the end of this list.

Adar, E. (1984) Quantification of aquifer recharge distribution using environmental isotopes and regional hydrochemistry. Ph.D. Thesis, Dept. of Hydrology and Water Resources, University of Arizona, Tucson.
Allison, G.B. and Hughes M.W. (1975) The use of environmental tritium to estimate recharge to a South-Australian aquifer. *J. of Hydrology* **26**, 245-254.
Andersen, L.J., Kelstrup, V., and Kristiansen, H. (1980) Chemical profiles in the Karup water-table aquifer, Denmark, In: *Nuclear Techniques in Groundwater Pollution Research*, IAEA, Vienna, 47-60.
Andrews, C.B. and Anderson, M.P. (1978) Impact of a power plant on the groundwater system of a wetland. *Groundwater* **16**, 105-111.
Andrews, J.N. and Lee, D.J. (1979) Inert gases in groundwater from the Bunter Sandstone of England as indictors of age and paleoclimatic trends. *J. of Hydrology* **41**, 233-252.
Andrews, J.N., Balderer, W., Bath, A.H., Clausen, H.B., Evans, G.V., Florkowski, T., Goldbrunner, J.E. Ivanovich, M., and Loosli, H. (1984) Environmental isotope study in two aquifer systems. In: *Isotope Hydrology 1983*, IAEA, Vienna 535-576.
Arad, A., Kafri, U., Halicz, L., and Brenner, I. (1984) Chemical composition of some trace and minor elements in natural groundwaters in Israel. *Geol. Survey Israel, Rep.* 29/84.
Back, W. and Hanshaw, B.B. (1970) Comparison of chemical hydrogeology of the carbonate peninsulas of Florida and Yucatan. *J. of Hydrology* **10**, 330-368.
Bath, A.H., Edmunds, W.M., and Andrews, J.N. (1979) Paleoclimatic trends deduced from the hydrochemistry of a Triassic sandstone aquifer, United Kingdom. In: *Isotope Hydrology, 1978*, IAEA, Vienna, **2**, 545-568.
Bebout, D.G. and Gutierrez, D.R. (1981) Geopressured geothermal resources in Texas and Louisiana - geological constraints. *Proc. 5th Conf. Geopressured-Geothermal Energy*, Baton Rouge, Louisiana, 13-24.
Bentley, H.W., and Davis S.N. (1980) Isotope geochemistry as a tool for determining regional groundwater flow. *Proc. 1980 National Terminal Storage Program Information Meeting*, Columbus, Ohio, 35-41.
Borman, F.H. and Likens, G.E. (1970) The nutrient cycles of an ecosystem. *Sci. A.* **223**(4), 92-101.

Borole, D.V., Gupta, S.K., Krishnaswami, S., Datta, P.S., and Desai, B.I. (1979) Uranium isotopic investigations and radiocarbon measurements of river-groundwater systems, Sabaramati Basin, Gujarat, India. In: *Isotope Hydrology 1978*, IAEA, Vienna, 1, 181-201.

Bortolami, G.C., Ricci, B., Suzella, G.F. and Zuppi, G.M. (1978). Isotope hydrology of the Val Coraoglia, Maritime Alps, Piedmont, Italy. In *Isotope Hydrology*, IAEA, Vienna, 327-350.

* Brassington, R. (1988) *Field Hydrology*. Open University Press, Milton Keynes; Halsted Press John Wiley, New York.

Bredenkamp, D.B., Schutte, J.M. and Du Toit, G.J. (1974) Recharge of a dolomitic aquifer as determined from tritium profiles. In: *Isotope Techniques in Groundwater Hydrology, 1974, Proc. Symp*, IAEA, Vienna, 1, 73-96.

* Carter, M.W., Moghissi A.A. and Kahn B. (eds) (1979) *Management of Low-Level Radioactive Wastes*. Pergamon Press, Oxford.

Castany, G. (1960) Quelques aspects nouveaux de l'hydrogeologie du basin parisien. *C.R. ann. Comité Fr. Geol. Geoph.*

Cherkauer, D.S. (1980) The effect of flyash disposal on a shallow groundwater system. *Ground Water* 18, 544-550.

Conrad, G., Jouzel, J., Merlivat, L., and Puyoo, S. (1979): La nappe de la craie en Haute-Normandie (France) et ses relations avec les eaux superficielles. In: *Isotope Hydrology 1978*, IAEA, Vienna, 1, 265-287.

Cotecchia, V., Tuzioli, G.S. and Magri, G. (1974) Isotopic measurements in research on seawater ingression in the carbonate aquifer of the Salentine Peninsula, Southern Italy. In: *Isotope Techniques in Groundwater Hydrology 1974*, IAEA, Vienna, 1, 445-463.

Craig, H. (1961a) Isotopic variations in meteoric waters. *Science* 133, 1702-1703.

Craig, H. (1961b) Standard for reporting concentrations of deuterium and oxygen-18 in natural waters. *Science* 133, 1833-1834.

Dansgaard, W. (1964) Stable isotopes in precipitation. *Tellus* 16. 436-469.

* Davis, S. and DeWiest, R.J.M. (1966) *Hydrogeology*. John Wiley, New York.

Davis, S.N. and Bentley, H.W. (1982): Dating groundwater, a short review. In: *Nuclear and Chemical Dating Techniques: Interpreting the Environmental Record*; Currie L.A. (ed) American Chemical Society Symposium Series 176, 187-222.

Deák, J. (1979) Environmental isotopes and water chemical studies for groundwater research in Hungary. In: *Isotope Hydrology 1978*, IAEA, Vienna, 1, 221-249.

De Marsily, G. (1986) *Quantitative Hydrology*. Academic Press, New York.

Dennis, H.W. (1973) Salt pollution of a shallow aquifer, Indianapolis, Indiana. *Ground Water* 11, 18-22.

Dowing, R.A., Smith, D.B., Pearson, F.J., Monksouse, R.A., and Otlet, R.L. (1977) The age of groundwater and its relevance to the flow mechanism. *J. of Hydrology* 33, 201-216.

* Drever, J.I. (1982) *The Geochemistry of Natural Waters*. Prentice-Hall, Englewood Cliffs, NJ.

* Eriksson, E. (1985) *Principles and Application of Hydrochemistry*. Chapman and Hall, London.

Evin, J. and Vuillaume, Y. (1970) Etude par le radiocarbone de la Nappecaptive de L'Albien du Bassin de Paris. In: *Isotope Hydrology 1970*, IAEA, Vienna, 315-332.

Fontes, J.Ch. (1983) *Dating of groundwater. Guide on Nuclear Techniques in Hydrology*; Technical Reports Series no. 91. IAEA, Vienna, 285.

Fontes, J. Ch., Bortolami, G.C. and Zuppi, G.M. (1979) Isotope hydrology of the Mont Blanc Massif. In: *Isotope Hydrology 1978*, IAEA, Vienna, 411-436.

Foster, K.E. and Fogel, M.M. (1973) Mathematical modelling of soil temperature. *Progressive Agriculture in Arizona* **25**, 10-12.

* Freeze, R.D. and Cherry, A. (1979) *Groundwater*. Prentice-Hall, Englewood Cliffs, NJ.

* Fritz P. and Fontes, J. Ch. (eds) (1980 and 1986) *Handbook of Environmental Isotope Geochemistry*. Vols 1 and 2.

Fritz, P., Drimmie, R.J., and Render, F.W. (1974) Stable isotope contents of a major prairie aquifer in central Manitoba, Canada. In: *Isotope Techniques in Groundwater Hydrology 1974*, IAEA, Vienna, **1**, 379-398.

Fritz, P. Matthess, G., and Brown, R.M. (1976) Deuterium and oxygen-18 as indicators of leachwater movement from a sanitary landfill. In: *Interpretation of Environmental Isotope and Hydrochemical Data in Groundwater Hydrology*, IAEA, Vienna, 131-142.

Fritz, P. Hennings, C.S., Suzulo, O., and Salati, E. (1979): Isotope hydrology in northern Chile. In *Isotope Hydrology 1978*, IAEA, Vienna, **2**, 525-544.

Fryberger, J.F. (1975) Investigation and rehabilitation of a brine contaminated aquifer. *Ground Water* **13**, 155-160.

Gat, J.R. (1971) Comments on the stable isotope method in regional groundwater investigations. *Water Resources Research* **7**, 980-993.

Gat, J.R. and Damsgaard, W. (1972) Stable isotope survey of the fresh water occurences in Israel and the Northern Jordan Rift Valley. *J. of Hydrology* **16**, 177-212.

Gat, J.R., Mazor, E. and Tzur, Y. (1969) The Stable isotope composition of mineral waters in the Jordan Rift Valley, Israel. *J of Hydrology* **7**, 334-352.

Geyh, M.A. (1972) Basic studies in hydrology and ^{14}C and ^{3}H measurements. *24th Int. Geol Cong.*, Montreal, **2**, 227-234.

Geyh, M.A., (1980) Interpretation of environmental isotopic groundwater data, arid and semi-arid zones. In: *Aridzone Hydrology: Investigations with Isotope Techniques*, IAEA, Vienna, 31-46.

Geyh, M.A. and Wirth, K. (1980) ^{14}C ages of confined groundwater from the Gwandu aquifer, Sokoto basin, Northern Nigeria. *J. of Hydrology* **48**, 281-288.

Goldenberg, L.C., Magaritz, M., and Mandel, S. (1983) Experimental investigation of irresvesible changes of hydraulic conductivity on the seawater-freshwater interface in coastal aquifers. *Water Resources Research* **19**, 77-85.

Gonfiantini, R., Dincer, T., and Derekoy, A.M. (1974) Environmental isotope hydrology in the Honda region, Algeria. In *Isotope Techniques in Groundwater Hydrology 1974*, IAEA, Vienna, **1**, 293-316.

Goolsby, D.A. (1971) Hydrochemical effects of injecting wastes into a limestone aquifer near Pensacola, Florida. *Ground Water* **9**, 13-19.

Heaton, T.H.E. and Vogel, J.C. (1981) 'Excess air' in groundwater. *J. of Hydrology* **50**, 201-216.

* Hem, J.D. (1985) *Study and Interpretation of the Chemical Characteristics of Natural Water*. U.S. Geological Survey, Water Supply Paper **254**.

Herzberg, O. and Mazor, E. (1979) Hydrological applications of noble gases and temperature measurements in underground water systems, examples from Israel. *J. of Hydrology* **41**, 217-231.

Hufen, T.H., Lau, L.S., and Buddemeier, R.W. (1974) Radiocarbon, ^{13}C and tritium in water samples from basaltic aquifers and carbonate aquifers on the island of Oahu, Hawaii. In *Isotope Techniques in Groundwater Hydrology 1974*, IAEA, Vienna, **2**, 111-127.

* Hutton, L.G. (1983) *Field Testing of Water in Developing Countries*. Water Research Centre, Medmeham, England.

Jorgensen, D.G. (1968) An aquifer test used to investigate a quality of water anomaly. *Ground Water* **6**, 18-20.
Kroitoru, L. (1987) The characterization of flow systems in carbonatic rocks defined by the groundwater parameters: Central Israel. Ph.D. Thesis, Feinberg Graduate School of the Weizmann Institute of Science, Rehovot, Israel.
Kroitoru, L., Carmi, I., and Mazor, E. (1987) Groundwater ^{14}C activity as affected by initial water-rock interactions in a carbonatic terrain with deep water tables: Judean Mountains, Israel. *Int. Symp. on the Use of Isotope Techniques in Water Resources Development*, IAEA, Vienna, extended abstract, 134-136.
Leontiadis, I.L., Payne, B.R., Letsios, A., Papagianni, N., Kakarelis, D., and Chadjiagorakis, D. (1983) Isotope hydrology study of Kato Nevroko of Dramas. In: *Isotope Hydrology 1983*, IAEA, Vienna, 193-206.
Levy, Y. (1987) The Dead Sea, hydrographic, geochemical and sedimentological changes during the last 25 years (1959-1984). *Geological Survey of Israel* (in Hebrew).
Lewallen, M.J. (1971) Pesticide contamination of a shallow bored well in southeastern Coastal Plains. *Ground Water* **9**, 45-49.
Loosli, H.H. and Oeschger, H, (1978) ^{39}Ar, ^{14}C and ^{85}Kr measurements in groundwater samples. In: *Isotope Hydrology 1978*. IAEA, Vienna, 2, 931-997.
* Matthes, G. (1982) *The properties of Groundwater*. John Wiley, New York.
Mazor, E. (1972) Paleotemperatures and other hydrological parameters deduced from noble gases dissolved in groundwaters: Jordan Rift Valley, Israel. *Geochimica et Cosmochimica Acta* **36**, 1321-1336.
Mazor, E. (1975) Atmospheric and radiogenic noble gases in thermal waters: their potential application to prospecting and steam production studies. *Proc. 2nd UN Symp. on Development and Use of Geothermal Resources*, San Francisco, **1**, 793-801.
Mazor, E. (1976) Multitracing and multisampling in hydrological studies. In: *Interpretation of Environmental Isotope and Hydrochemical Data in Groundwater Hydrology*, IAEA, Vienna, 7-36.
Mazor, E. (1978) Mineral waters of the Kinneret basin and possible origin. In: *A Monography on Lake Kinneret*, Serruya, C. (ed.) W. Junk N.V. Publishers; The Hague 103-120.
Mazor, E. (1979) Dilute water-rock reactions in shallow aquifers of the Kalahari flatland. *Proc. 3rd Water-Rock Interaction Symp.*, Edmonton, Canada, 14-15.
Mazor, E. (1982) Rain recharge in the Kalahari — a note on some approaches to the problem. *J. of Hydrology* **55**, 137-144.
Mazor, E. and Bosch, A. (in press). He as a tool for groundwater dating in the range of 10^4 to 10^8 years. In: *Isotopes of Noble Gases as Tracers in Environmental Studies*. IAEA, Vienna.
Mazor, E. and Verhagen, B. Th. (1983) Dissolved ions, stable isotopes and radioactive isotopes and noble gases in thermal waters of South Africa. *J. of Hydrology* **63**, 315-329.
Mazor, E. and Mero, F. (1969a) Geochemical tracing of mineral and fresh water sources in the Lake Tiberias basin, Israel. *J. of Hydrology* **7**, 276-317.
Mazor, E. and Mero. F. (1969b) The origin of the Tiberias-Noit mineral water association in the Tiberias - Dead Sea Rift Valley, Israel. *J. of Hydrology* **7**, 318-333.
Mazor, E. and Kroitoru, L. (1987) Phreatic-confined discontinuities and restricted flow in confined groundwater systems. *Int. Symp. on the Use of Isotope Techniques in Water Resources Development*, IAEA, Vienna, extended abstract, 130-131.

Mazor, E., Kaufman, A., and Carmi, I, (1973) Hammet Gader (Israel): Geochemistry of a mixed thermal spring complex. *J. of Hydrology* **18**, 289-303.

Mazor, E., Nadler, A., and Harpaz, Y. (1973) Notes on the geochemical tracing of the Kaneh-Samar spring complex, Dead Sea basin. *Israel J. of Earth Sciences* **22**, 255-262.

Mazor, E., Rosenthal, E. and Eckstein, J. (1969). Geochemical tracing of mineral water sources in the South-Western Dead Sea Basin, Israel. *J. of Hydrology*, **7**, 246-275.

Mazor, E., Verhagen, B. Th., and Negrenenu, E. (1974) Hot springs of the igneous terrain of Swaiziland. In *Isotope Techniques in Groundwater Hydrology 1974*, IAEA, Vienna, **2**, 29-47.

Mazor, E., Verhagen, B. Th., Sellschop, J.P.F., Jones, M.T., Robins, N.E., Hutton, L., and Jennings, C.M.H. (1977) Northern Kalahari groundwaters: hydrologic, isotopic and chemical studies at Orapa, Botswana. *J. of Hydrology* **34**, 203-234.

Mazor, E., Verhagen, B. Th., Sellschop, J.P.F., Jones, M.T., and Hutton, L.C. (1981) Sodium exchange in a NaCl waste disposal case (Lobatse, Botswana): implications to mineral water studies. *Environ. Geol.* **3**, 195-199.

Mazor, E., Vautaz, F.D., and Jaffé, F.C. (1985) Tracing groundwater components by chemical, isotopic and physical parameters, example: Schinziach, Switzerland. *J. of Hydrology* **76**, 233-246.

* Ozima, M. and Podosek, F.A. (1983) *Noble Gas Geochemistry*. Cambridge University Press.

Payne, B.R. and Yurtsever, Y. (1974) Environmental isotopes as a hydrogeological tool in Nicaragua. In *Isotope Techniques in Groundwater Hydrology 1974*, IAEA, Vienna, **1**, 193-202.

Payne, B.R., Quijano, L., and Latorred, C.D. (1980) Study of the leakage between two aquifers in Hermosillo, Mexico, using environmental isotopes. In: *Arid Zone Hydrology: Investigations with Isotope Techniques*, IAEA, Vienna, 113-130.

Pearson, Jr., F.J. and Hanshaw, B.B. (1970) Sources of dissolved carbonate species in groundwater and their effects on carbon-14 dating. In: *Isotope Hydrology 1970*, IAEA, Vienna, 271-286.

Pearson, F.J. and Swarzenski, W.V. (1974) ^{14}C evidence for the origin of arid region groundwater, northeastern province, Kenya. In: *Isotope Techniques in Groundwater Hydrology 1974*, IAEA, Vienna, **2**, 95-109.

Pearson, G.N. (1974): Tritium data from groundwater in the Kristianstad Plain, Southern Sweden. In: *Isotope Techniques in Groundwater Hydrology 1974*, IAEA, Vienna, **1**, 45-56.

Pickens, J.F., Cherry, J.A., Grisak, G.E., Merritt, W.F., and Risto, B.A. (1978) A multilevel device for ground-water sampling and piezometric monitoring. *Ground Water* **16**, 322-327.

Rafter, T.A. (1974) The dating of fossil man in Australia. *Proc. Symp. Hydrogeochemistry and Biogeochemistry*. Tokyo, Japan.

Rudolph, J., Rath, H.K., and Sonntag, C. (1983) Noble gases and stable isotopes in ^{14}C-dated paleowaters from central Europe and the Sahara. In: *Isotope Hydrology 1983*, IAEA, Vienna, 467-477.

Saffinga, P.G. and Keeney, D.R. 61977) Nitrate and chloride in ground water irrigated agriculture in Central Wisconsin. *Ground Water* **15**, 170-177.

Salati, E., Matsui, E., Leal, J.M., and Fritz, P. (1980) Utilization of natural isotopes in the study of salinization of the waters in the Pajeu River Valley, Northeast Brazil. In: *Arid-Zone Hydrology: Investigations with Isotope Techniques*, IAEA, Vienna, 133-151.

Schoch-Fischer, H., Rozanski, K., Jacob, H.J., Sonntag, C., Jouzel, I., Ostlund, G., and Geyh, M.A. (1983) Hydrometeorological factors controlling the time variation of D, ^{18}O and ^3H in atmospheric water vapour and precipitation in the northern westwind belt. In: *Isotope Hydrology 1983*, IAEA, Vienna, 3-30.

* Schoeller, H. (1954) *Arid Zone Hydrology — Recent Developments*. UNESCO, Paris.

Shampine, W.J., Dincer, T., and Noory, M. (1979) An evaluation of isotope concentrations in the groundwater of Saudi Arabia. In: *Isotope Hydrology 1978* IAEA, Vienna, 2, 443-463.

Shuster, E.T. and White, W.B. (1971) Seasonal fluctuations in the chemistry of limestone springs: a possible means for characterizing carbonate aquifers. *J. of Hydrology* **14**, 93-128.

Siegenthaler, U. and Oeschger, H. (1980) Correlation of ^{18}O in precipitation with temperature and altitude. *Nature* **285**, 314-317.

Sonntag, C., Klitzsch, E., Lohnert, E.P., Ee-Shazly, E.M., Munnich, K.O., Junghans, Ch., Thorweihe, U., Weistroffer, K. and Swailem, F.M. (1979) Paleoclimatic information from deuterium and oxygen-18 in carbon-14 dated north Saharian groundwaters. In: *Isotope Hydrology 1978*, IAEA, Vienna, 2, 569-581.

* Stahl, W., Aust, H. and Dounas, A. (1974) Origin of artesian and thermal waters determined by oxygen, hydrogen and carbon isotope analyses of water samples from the Sperkhios Valley, Greece. In: *Isotope Techniques in Groundwater Hydrology 1974*, IAEA, Vienna, 1, 317-339.

Tamers, M.A. and Scharpenseel, H.W. (1970) Sequential sampling of radiocarbon in groundwater. In: *Isotope Hydrology 1970*, IAEA, Vienna, 241-257.

Tremblay, J.J., D'Cruz, J., and Anger, H. (1973) Salt water intrusion in the Summerside area, P.E.I. *Ground Water* **11**, 21-27.

* Todd, D.K. (1980) *Groundwater Hydrology*. Second edition, John Wiley, New York.

Verhagen, B.Th., Mazor, E. and Sellschop, J.P.F. (1974) Radiocarbon and tritium evidence for direct recharge to groundwaters in the Northern Kalahari. *Nature* **249**, 643-644.

Verhagen, B.Th., Smith, P.E., McGeorge, I., and Dzimebowski, Z. (1979) Tritium profiles in Kalahari sands as a measure of rain-water recharge. In *Isotope Hydrology 1978*, IAEA, Vienna, 2, 733-751.

Vogel, J.C. (1970) Carbon-14 dating of groundwater. In: *Isotope Hydrology 1970*, IAEA, Vienna, 225-239.

Vogel, J.C. and Van Urk, H. (1975). Isotopic composition of groundwater in semi-arid regions of Southern Africa. *J. of Hydrology*, **25**, 23-36.

Vogel, J.C., Thilo, L., and Van Dijken, M. (1974) Determination of groundwater recharge with tritium. *J. of Hydrology* **23**, 131-140.

Vuataz, F.D. (1982) Hydrologie, géochimie et géothermie des eaux thermales de Suisse et des regions Alpines limitrophes. *Matériaux pour la géologie de la Suisse - hydrologie* No. 29. Kummerly & Frey Geographischer Verlag, Berne.

* Walton, W.C. (1988) *Groundwater Pumping Tests Design and Analysis*. H.K. Lewis, London.

Wilmoth, B.M. (1972) Salty groundwater and meteoric flushing of contaminated aquifers in West Virginia. *Ground Water* **10**, 99-105.

Wilson, L.G. and DeCook, K.J. (1968) Field observations on changes in the subsurface water regime during influent seepage in the Santa Cruz River. *Water Resources Research* **4**, 1219-1233.

Winslow, J.D., Stewart, Jr., H.G., Johnston, R.H. and Crain, L.J. (1965) Groundwater resources of eastern Schenectudy County, New York, with emphasis on infiltration from the Mohawk River. State of New York Conservation Department Water Resources Commission Bull. **57**, 148.

Yurtsever, Y. and Payne, B.R. (1979) Application of environmental istotopes to groundwater investigation in Qatar. In: *Isotope Hydrology 1978*, IAEA, Vienna, **2**, 465-490.

Special publications of the International Atomic Energy Agency P.O. Box 100, Vienna, Austria.

Application of Isotope Techniques in Hydrology, 31 pp., 1962.
Arid-Zone Hydrology: Investigations with Isotope Techniques, 265 pp., 1980.
Concentration in Precipitation (1969-1983), Vols 1-7.
Environmental Isotope Data No. 1: World Survey of Isotope Concentration in Precipitation 1969-1983 Vols 1-7.
Interpretation of Environmental Isotope and Hydrochemical Data in Groundwater Hydrology, 228 pp., 1976.
Isotopes in Hydrology, 740 pp., 1967.
Isotope Hydrology, 1978, Vol. 1: 440 pp., Vol. II: 540 pp., 1979.
Isotope Hydrology 1983, 874 pp., 1984.
Isotope Techniques in Hydrology, (Bibliography) Vol. I: 1857-1965, 228 pp., 1968; Vol. II: 1968-1971, 233 pp., 1973.
Isotope Techniques in the Hydrological Assessment of Potential Sites for the Disposal of High-Level Radioactive Wastes, 164 pp., 1983.
Isotope Techniques in the Study of the Hydrology of Fractured and Fissured Rocks; 306 pp., 1986.
Nuclear Techniques in Groundwater Pollution Research, 286 pp., 1980.
Paleoclimates and Paleowaters: A Collection of Environmental Isotope Studies, 216 pp., 1983.
Stable Isotope Hydrology: Deuterium and Oxygen-18 in the Water Cycle; 339 pp., 1981.
Statistical Treatment of Environmental Isotope Data in Precipitation, 276 pp., 1981.
Tritium and Other Environmental Isotopes in the Hydrological Cycle, 83 pp., 1967.

INDEX

Abattoir waste 247-252
Abundances, isotopic 59
Accuracy 63
Aerated zone 9
Age, effective 148
Algeria 139-140
Alkalinity 118
Altitude effect 133-137, 142-145
Amount effect 130-131
Anion 60
Annual recharge calculation 155
Appalachians, U.S.A. 47
Aquiclude 13
Aquifer 12
Aravaipa Valley, Arizona 44-45
Artesian flow 15
Atomic weight 58-59
Austria 12

Blumau aquifer, Austria 192
Bossa Cave, Italy 137
Botswana 99-100, 125, 153, 166-167
Brazil 127, 133, 139
Bunter sandstone aquifer,
    England 186-191, 209-211

Canada 95-96, 128
Carbon-13, applied to C-14,
    dating 172-174
  in groundwater 170-172
Carbon-14, groundwater
    dating 166-175
  indicating mixing 193-196
  manmade dilution 165
  natural production 164
  nuclear bombs 166
Carbon isotopes 58, 164
Cation 60

Chad basin 168-169
Charlston, Virginia, U.S.A. 95-96
Checks of laboratory 65
Chile 128
Chlorinity 94
Chott-el-Honda, Algeria 139-140
Clay 13, 24
CO_2 9, 91-91
Coal, ash contamination 237-241
Combioula, Switzerland 100-101
Composition diagrams 79-82, 84, 87
Concentration units 61-62
Conduits 10-11, 16-17, 47, 155, 197
Conductivity 18, 26
Confined aquifers 15
Conglomerate 12
Continental effect 132

Darcy's law 18-19
Dating, C-14 166-175
  helium-4 191-192, 214-216
  tritium 151
Dead Sea, Israel 62, 88-90
Decay curve, carbon-14: 165
  tritium 147-148
Depth of circulation 45-47, 51, 205, 211-212
Depth profiles 105-106, 220-221
Discharge 22, 53-56

Efflorescences 120-121
Electrical conductance 115-116
Elements, definition 57
End members, mixing
    groundwaters 89-91, 159-160
England 186-191, 209-211
Equipment list, field work 121
Equivalence units 61-62

Errors, analytical 64–65
Evaporation, detection 20, 123, 139–140
  line 123, 140
Evaporites 24
Evapotranspiration 20

Faults influencing flow 29–30
Fingerprint diagrams 74–78
Fissures, flow 10, 26
Flow direction 40–42, 82–84
Florida limestone aquifer, U.S.A. 192, 225–230
Fluid waste injection 225–230
Folded structures 27–29
'Forbidden' combinations 90–91, 118, 193–196
Fractionation, isotopic 123–125
France 132, 160–163, 175–181

Geothermal, gradient 45–46
  systems 52
Gradient, water level 40–42
Greece 142–145
Gypsum 24

Half life 147–148
Hamei Yesha springs, Israel 124
Hamei Zohar springs, Israel 124
Hammat Gader springs, Israel 193–196
Hawaii, U.S.A. 166–169
Helium, dating 189–191, 214–216
  radiogenic 214–216
Hermosillo, Gulf of California, Mexico 142
Historical data 107–108, 218
Hungary 197
Hydraulic, discontinuity 184–193
  interconnection 40–41, 82–84
  pulse propagation 39
Hydrogen, isotopes 58
  isotopic abundances in sea water 59
Hydrograph 35

Igneous rocks 12
Industrial fluid injection 225–230
Infiltration 9–11, 25, 26
India 181–184
International Atomic Energy Agency (IAEA) 149–150. 270
Intrusive bodys 30

Ionian Sea coast, Italy 97
Ions 59–60
Isotopes, composition units 122–123
  definition 57
  of water 122
Isotopic fractionation 123–125
Israel 62, 87–90, 104–105, 124, 145–146, 151, 184–186, 194–196, 201–205
Italy 97, 136–137, 140–141, 160–163

Judean Mountains, Israel 184–186

Kalahari, Botswana 99–100, 153, 166–168
Kaneh-Samar springs, Israel 88–90
Karstic, recharge 213–214
  systems 16–17, 47

Limestone 12
Lincolnshire limestone aquifer, England 191
Lithological controls 23–26
Louisiana, USA 46

Maritime Alps, Italy 136–137
Meteoric line 125–129, 144
Mexico 142
Milliequivalent units 62
Mixing, of groundwaters 54–56, 80–81, 85–91, 124, 140–142, 145, 159, 193–196
  percentages 85–88, 145
Mohawk River, N.Y., U.S.A. 38–40, 48–50, 156, 158
Mont Blanc tunnel 160–163

Nicaragua 134–135
Nitrate contamination 220–225
Noble gases, properties 60, 198–216
  altitude-solubility correction factors 199–200
  atmospheric concentrations 198
  atmospheric inputs 199
  atmospheric isotopic abundances 199
  paleotemperatures 209–211
  percent retention 213–214
  recharge intake temperatures 201–205
  solubilities 199

Oil brines, contamination 232–236

Index

Orapa, Botswana 125
Oxygen, dissolved 117–118
  isotopes 58
  isotopic abundances in sea water 59

Paj'eu River, Brazil 127, 139
Paleowater 127–129
Paleotemperatures 209–211
Paris basin, France 175–181
Permeability 18, 27
Pesticide contamination 225
pH 116
Phreatic aquifer 14
Piezometric water head 15
Piston flow 10, 153, 197, 222–223
Planning list 111–112
Pollution, monitoring 217–218
  tracing principles 217–218
  the aerated zone 219

Qatar 123, 158–159
Quality, of data 63–66
  of water 23–25

Radioactive decay curve,
    carbon-14 165
  tritium 147–148
Reaction error 64–65, 70–71
Recharge 21, 25–26
Recharge, along vertical
    joints 160–163
  altitude, isotopic 133–137, 142–145, 163
  areas 21, 212–213
  front, travel 39
  indications 155
  modes 37–39
Reconnaissance studies 111
Repeated measurements (time
    series) 95, 108–109
Reproducibility 63, 67, 70
Resolution, analytical 63
Road salting contamination 247–248

Salentine Peninsula, Italy 140–141
Sampling 102–107, 119–120
Sandstone 12
Sanitary landfill, pollution 241–245
Saratoga National Historic Park,
    N.Y. 37
Saturated zone 11
Sea water intrusion 140, 142, 145

Shale 14, 24
Sigma, error 69
Significant figures 64
Sill 30–31
SMOW (Standard Mean Ocean
    Water) 122–123
Soil zone 9
South Africa 138, 154–156, 174–175
Sperkhios Valley, Greece 142–145
Storage, water 12, 26
Structural control 27–31
Sweden 145
Switzerland 55–56, 97–99, 100–101,
    129–130, 133, 134, 160

Tables, arrangement 66–70, 72–73
TDI (total dissolved solid ions) 69, 94
TDS (total dissolved solids) 94
Temperature, effect 129–130
  groundwater tracing 48–52
  measurement 45, 115
  tracing cooling water 236–237
Temperatures of the hydrological
    cycle 202–205
Thermal gradient 46
Time-data series 95, 108–109
Transects 36–37, 119
Transpiration 21
Transval, South Africa 154
Tritium, properties 147
  ages 151–153, 160–163
  nuclear bomb production 148–152
  indicating mixing 159–160, 193–194
  indicating recharge 156–159
  natural production 147–151
  soil profiles 154–157
  unit 147

Valences 58–60, 69
Vals, Switzerland 159–160
Velocity of flow 18, 42, 174–175, 183
Vienna, Austria 151

U.S.A. 37–40, 46, 48–50, 95–96, 151,
    168, 192, 212
UT (tritium unit) 147

Warm springs 138
Water cycle 2–4
  encoded information 2–4
  table 32–34
  warm 50–52

Water-rock interactions 23–35, 91–93, 169–171
Well design 25

Yverdon, Switzerland 55–56